高等院校土建学科双语教材（中英文对照）

◆ 土木工程专业 ◆

BASICS

# 玻璃构造

# GLASS CONSTRUCTION

[德] 安德烈斯·艾奇里斯
黛安·娜维拉蒂尔 著

袁 磊 许雪松 译

中国建筑工业出版社

著作权合同登记图字：01－2009－7705号

**图书在版编目（CIP）数据**

玻璃构造/（德）艾奇里斯，娜维拉蒂尔著；袁磊，许雪松译．—北京：中国建筑工业出版社，2011.2
高等院校土建学科双语教材（中英文对照）◆土木工程专业◆
ISBN 978－7－112－12846－4

Ⅰ.①玻… Ⅱ.①艾…②娜…③袁…④许… Ⅲ.①玻璃结构－高等学校－教材－汉、英 Ⅳ.①TU382

中国版本图书馆CIP数据核字（2011）第006451号



责任编辑：孙　炼
责任设计：陈　旭
责任校对：陈晶晶　姜小莲

高等院校土建学科双语教材（中英文对照）
◆土木工程专业◆
**玻璃构造**
［德］安德烈斯·艾奇里斯　黛安·娜维拉蒂尔　著
袁　磊　许雪松　译
*
中国建筑工业出版社出版、发行（北京西郊百万庄）
各地新华书店、建筑书店经销
北京嘉泰利德公司制版
北京云浩印刷有限责任公司印刷
*
开本：880×1230毫米　1/32　印张：4½　字数：144千字
2011年5月第一版　2011年5月第一次印刷
定价：**15.00**元
ISBN 978－7－112－12846－4
（20119）

# 中文部分目录

## CONTENTS

# 序

玻璃是最具魅力的建筑材料之一。它既连结空间又分割空间。玻璃种类繁多，从令人一目了然的全透明玻璃到隔绝外界干扰的立面反射玻璃，不一而足。玻璃的多样性使其在建筑设计中独树一帜。

玻璃形式纯净，却需要仔细斟酌。它可能意外地突然破碎，也不耐机械压力。不过，得益于科研的不断进展，玻璃展现出巨大的发展潜力及多样的应用可能——包括防弹玻璃和全玻璃承重结构之类的创新性技术，在这两方面其他建材望尘莫及。

一直以来，建材玻璃应用的发展都伴随着对其技术特性和可能性的认识的发展。建筑师只有了解各种玻璃的特性、玻璃构造的组成和组件及其材料局限性，才能开发出有创意的玻璃应用方法，并不断突破现有局限。

本书是国外高等院校土建学科基础教材系列丛书中的一个分册，从构造角度入手，向读者介绍玻璃的特殊性能及其作为建筑材料的应用可能。本书通过增进建材知识、解释复杂的构造和各种用途，使建筑学学生有能力构想具有个人创意的方案，而不仅仅是遵从建造行业已有的条条框框。玻璃构造的很多发展不只是从材料学实验室得来，而是来自建筑师富于创意、突破常规的设计，这些建筑师解决了难题，并促成了新的发展和应用。本书旨在启发学生运用玻璃方面的知识为设计增添创意，甚或开拓出新的创作思路。

**套书编辑 Bert Bielefeld**

# FOREWORD

Glass is one of the most attractive building materials. It connects spaces to one another while separating them. The various types of glass range from complete transparency and openness to reflective glass that provides hermetic separation for facades. This diversity makes glass unique in architectural design.

Glass in its pure form is a material that needs to be carefully considered. It breaks very quickly and often unexpectedly and is sensitive to mechanical stress. Thanks to advances in research, however, no other building material offers such great potential for development or such a multifaceted spectrum of possible uses—including bulletproof glass and the creation of bearing structures entirely of glass.

The use of glass as a building material is always coupled with knowledge of technical properties and possibilities. Only by knowing the properties of various glasses, the components and elements of a glass construction, and the limits of the material can an architect develop creative solutions using this material and transcend the existing limits again and again.

The present volume, *Glass Construction*, is part of a subset of the series on construction. It begins by considering construction and conveys to the reader an understanding of the specific properties of glass and the possibilities that it offers as a building material. By increasing knowledge of the building material and revealing complex structures and applications, it enables students of architecture to consider their own creative solutions beyond the standardized offerings of the construction industry. Many advances in glass construction have resulted not only from laboratory research on the material but also from innovative and unconventional designs by architects who have met the challenges and provided stimuli for ever new developments and uses. *Basics Glass Construction* is intended to inspire students to use their knowledge of glass to explore the possibilities for their own designs and perhaps even develop new approaches.

Bert Bielefeld, Editor

# INTRODUCTION

Like few other materials, glass possesses a symbolism that transcends mere function and exerts a particular fascination. The glass windows of the Gothic period already deliberately played with light in order to produce a feeling of transcendence. In the architectural visions of modernism, this transparent material took on central significance, although glass played different roles, depending on the theoretical approach in question. Glass was appreciated not only for its transparency, which permitted an almost dematerialized shell and hence flowing, open spaces, but also for its graceful, angular, and glittering qualities. The emancipation of glass from its role as a filler in relatively small windows to become an autonomous element would prove particularly forward-looking. Problems in terms of energy conservation and disregard of the physical requirements of construction put a temporary end to the euphoria over this material.

Today, thanks, among other things, to a turn to solutions that make sense for energy conservation, and the development of glass that provides effective insulation and solar control, glass has once again become a high-performance material. As such, it fulfills both functional and design requirements and opens up ever new areas of application.

For all the possibilities that glass offers, however, it should not be forgotten that it is a very brittle material. Glass breaks suddenly and without warning when it is overstressed in particular places. That calls for precise knowledge of the nature of the material and great care in planning and implementing glass structures.

This book introduces students step by step to the basics of glass as a building material and to glass construction. In the first three chapters, the reader learns the properties and diversity of today's types of glass, then the most important principles for constructing with it, and finally the different areas of application and their constraints. The technical fundamentals are explained intelligibly and in a structured way in terms of their principles and by means of simple examples. In this way, students obtain an overview of the current state of technology and are in a position to plan their own projects using glass as a building material and to make their own ideas reality.

# GLASS AS A BUILDING MATERIAL

## THE PRODUCTION OF GLASS

Composition

Glass is produced by heating a mixture that consists largely of silica (silicon dioxide) and soda ash (sodium carbonate). Soda ash serves as a so-called flux to reduce the high melting point of silica (approx. 1700 °C). The melting that then takes place above 1100 °C is amorphous—that is, virtually no crystals are formed. Because the structure of glass resembles that of fluids, glass is sometimes called a "supercooled liquid." › Tab. 1

Types of glass

The most commonly used glass in architecture is soda-lime glass, the main components of which are silicon dioxide, calcium oxide, and sodium oxide. Borosilicate glass, which contains boron oxide rather than calcium oxide, is often used as fire-resistant glass thanks to its high chemical and thermal stability. Lead glass, which is produced from lead crystal, among other things, and special glass, which is used in optical devices, for example, have no significance for architecture. Glass ceramic, by contrast, has recently begun to be used to clad facades. Transparent synthetic glass such as acrylic glass and polycarbonate is lighter and easier to work than mineral glass, but because it has a lower surface hardness it is considerably more sensitive to scratching and thus not as durable.

## BASIC PRODUCTS

Glass products that are formed during production in a "hot" state or immediately after cooling are generally referred to as basic products or basic glass. Various kinds of basic glass are employed in architecture. In addition to clear flat glasses with a smooth surface, glasses with specially designed surfaces or special shapes are also used. Basic glass is often further processed or finished. › Chapter Processing and finishing

Tab. 1:
Composition of glass according to EN 572, Part 1

| | | |
|---|---|---|
| Silicon dioxide | $SiO_2$ | 69-71% |
| Calcium oxide | $CaO$ | 5-12% |
| Sodium oxide | $Na_2O$ | 12-16% |
| Magnesium oxide | $MgO$ | 0-6% |
| Aluminum oxide | $Al_2O_3$ | 0-3% |

We describe below the types of basic glass relevant to architecture and how they are produced.

### Float glass

Float method

Float glass is the most commonly used form of basic glass. Its name derives from the process by which it is produced. The float method developed in 1960 was a milestone in the history of the production of flat glass, because it became possible for the first time to produce large quantities of clear, transparent glass with nearly flat surfaces.

Production begins by melting the raw materials, referred to as "batch," in a furnace. Next, the molten glass runs onto a flat bath of molten tin. Because it has a lower specific gravity, the glass floats on the tin, which gives it its flat surface. This produces an endless glass ribbon that slowly hardens; its thickness is determined by the speed with which it is pulled over the tin bath. After passing through this tin bath, or "float bath," the glass ribbon passes through a cooling zone, and finally is cut into plates. › Fig. 1

The standard format, known as ribbon size, is 600 × 321 cm. Its standard or nominal thicknesses are 2, 3, 4, 5, 6, 8, 10, 12, 15, and 19 mm.

### Sheet or window glass

The term "window glass" is somewhat misleading, since float glass is usually used for windows these days. Sheet or window glass is now produced only individually in drawn glass facilities, in which the glass ribbon is drawn horizontally or even vertically from the furnace. Today this process is used to produce particular kinds of colored glass, or for special

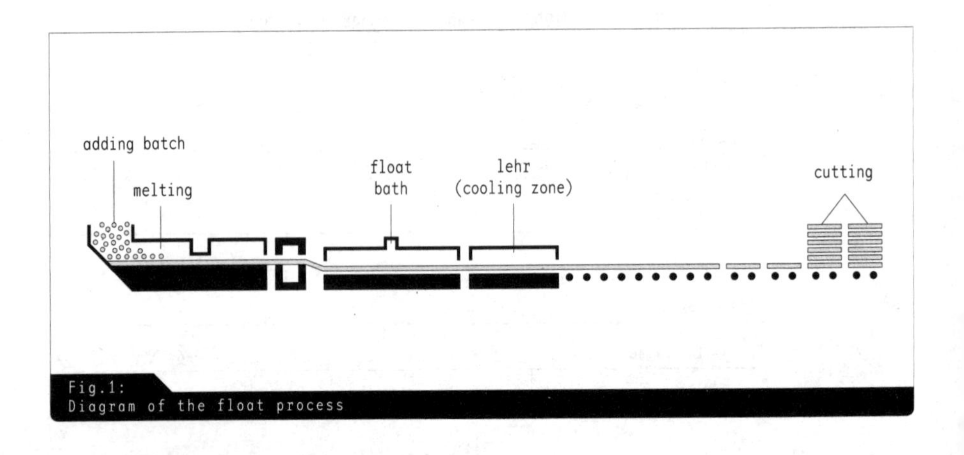


Fig. 1:
Diagram of the float process

Fig.2:
Ornamental glass

Fig.3:
Various types of profile glass

glass such as very thin glass. The surface quality is somewhat poorer than that of float glass, as waves (called striae) are visible.

### Cast glass

Rolling method

Cast or rolled glass is produced using the rolling method, in which the glass mass is formed into an endless ribbon between two water-cooled rollers. Decorations engraved into the rollers give the glass ribbon a surface structure. In order to produce wired glass or ornamental wired glass, a wire net can be rolled into the glass. Cast glass is also called ornamental glass because of its structure or ornamental surface, and its uses include partitions and facade openings where an open view through is neither desired nor required. › Chapter Design glass, Ornamental glass

### Profile glass

Profile glass is produced using a process similar to that for cast glass. In addition to a surface structure, the glass is given a cross section (U-profile) that is structurally advantageous, so that considerable spans become possible. › Fig. 3

Because it is economical, profile glass has been and continues to be used for the facades of industrial buildings. Nowadays, profile glass is also a very popular building material in architecture generally. › Chapter Applications, Profile glass

Fig.4:
Glass bricks

### Pressed glass

Glass bricks

Pressed glass is the general term for glass bricks, glass ceiling tiles, and concrete glass. These are made by fusing two bodies of glass pressed in forms (pressing method). When cooled, the air in the hollow space within the glass brick is under low pressure, which makes it nearly impossible for condensation to form. Glass bricks are often used in interiors or as translucent elements in solid exterior walls. Architectural glass is used primarily in ferroconcrete construction, since it is also well suited to higher structural loads. › Fig. 4

## PROCESSING AND FINISHING

Most types of basic glass are further processed and finished after manufacture, which offers an opportunity to influence not only its form and shape but also the physical and structural properties. The spectrum of finishing processes ranges from mechanical and heat treatments to coating and designing the surfaces.

### Mechanical processing

Processing such as cutting, boring, grinding, and polishing are generally labeled mechanical processing or mechanical finishing.

Cutting

Glass is cut into its desired shape. It is not really a cutting process, since the cutting wheel or diamond merely scratches the glass surface, and then the glass is broken by gently bending it along the scratched line. The glass coming directly from the floating machine is cut to ribbon size (600 × 321 cm) and is then given its desired final size in the finishing process. Cutting down and further processing are usually done by machine. For example, complex forms can be produced very precisely using a water jet cutter. The cutting is done with the aid of high-pressure jet of water (with

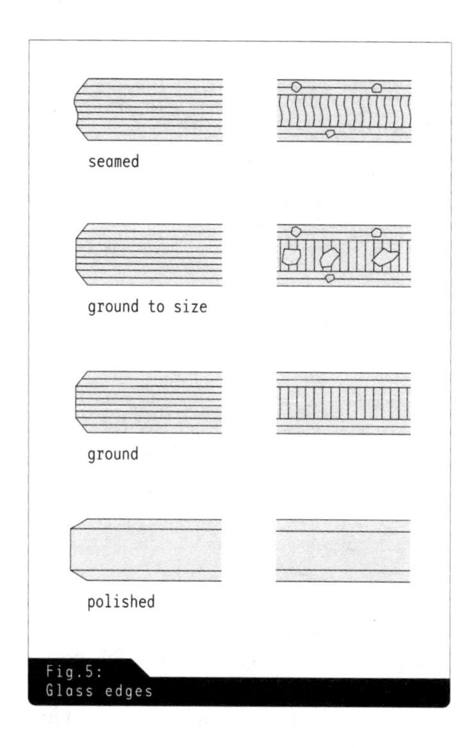


Fig.5:
Glass edges

a water pressure up to 6000 bar) to which a cutting agent (an abrasive) is added.

Edge treatments

Because the edges of the glass are still sharp after cutting, it is necessary to treat the edges both to prevent injury and for reasons relating to the production process. The treatments are distinguished in the following illustration and table. › Fig. 5 and Tab. 2

\\ Note:
The edge treatment affects not only the optical qualities of the glass plate but also its stability. Inhomogeneous or sharp edges increase the likelihood of damage to the glass (cracks, shells). The type of edge treatment must therefore be determined before awarding the contract; in some cases the pattern should as well.

Tab. 2:
Edge treatments

| Term | Definition |
|---|---|
| Cut | Untreated edge of glass with sharp perimeters as a result of cutting flat glass |
| Seamed | Cut edge with perimeters that have been smoothed with a grinding tool |
| Ground to size | Pane ground to the desired size. The edge may have shiny spots and shells. |
| Ground | The entire edge is ground to a semimatte finish. Shiny spots and shells are not permitted. |
| Polished | Ground edge is polished |

Boring

Various applications in modern glass construction call for boreholes within the pane of glass. These boreholes make it possible to fasten the plates of glass at these points. Because glass is a very hard and brittle material, any mechanical processing must be done with an appropriate tool—boring, for example, should be done with a diamond-tip, water-cooled hollow drill. It should be drilled from both sides at once to prevent breaking through the opposite side. Because local tensions can be very high around the inner face of the hole, the panes are tempered after drilling to increase the strength of the glass.

**Tempering**

Bent

Bent or curved glass is produced from flat glass that is heated to approx. 600 °C to soften it and then brought into the desired form. Glass can be bent along one axis (cylindrically) or two (spherically), as for an all-glass domed ceiling, for example.

Tempering

Tempered safety glass (TSG) is tempered glass that has been heated to approx. 600 °C under controlled conditions in a tempering furnace and then cooled very quickly. The state of tension in the glass that this process produces can be "frozen," which considerably increases the bending strength of the material. › Fig. 6

In addition, TSG has a much higher thermal shock resistance than float glass. TSG can resist a thermal shock of as much as 150 K, whereas float glass can only withstand thermal shock of 40 K. TSG is considered safety glass, however, primarily due to the way it breaks. Because it is in

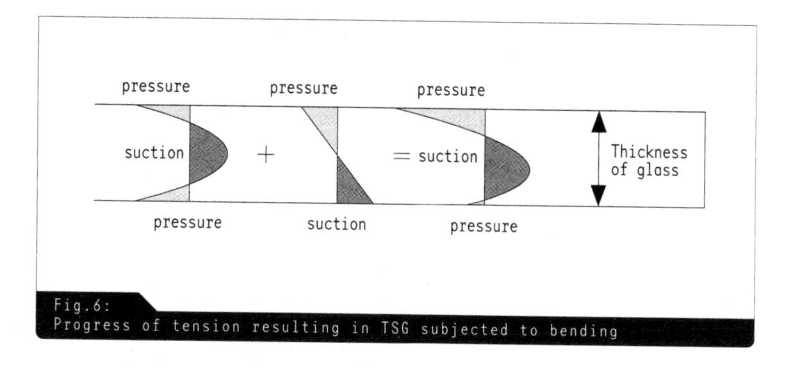


Fig. 6:
Progress of tension resulting in TSG subjected to bending

state of internal tension, when it breaks it shatters suddenly into small fragments with blunt edges, which considerably reduces the risk of severe injury. On the one hand, because the fragments remain hooked together, it has the advantage that the broken pane usually stays in the frame. On the other, the hooked fragments also pose a threat to people underneath the glazing, as relatively large continuous pieces can fall down.

> Test heat soak

A rare but nonetheless undesirable property of toughened glass is spontaneous breakage: tiny inclusions of nickel sulfide, invisible to the naked eye, expand in volume over time, and can cause the pane to break unexpectedly even years after installation. One reliable means of detecting nickel sulfide inclusions is the heat soak test, in which the panes are warmed to about 290 °C. During a soak time of about four hours in the heat soak oven, the TSG panes with nickel sulfide inclusion will probably break and hence never be installed in the first place. In accordance with EN 14179, heat-soaked TSG is given the standard label TSG-H.

\\Note:
Because TSG is itself under tension, it cannot be cut, bored, or ground subsequently, as the pane would break. All the necessary steps for the mechanical finishing must therefore be completed before it is tempered.

Under certain daylight conditions and under polarized light, anisotropies become visible in TSG that has developed a directional (anisotropic) structure in the tempering process. Caused by double refraction of light rays in the areas of tension, they reveal patterns or clouds of structure in the spectral colors.

Heat-strengthened glass

Heat-strengthened glass (HSG) is produced by a process similar to that for TSG. It is, however, cooled more slowly, which reduces the pressure on the surface, which results in both lower bending strength and a different fracture pattern than that of TSG. The fracture pattern of HSG resembles that of untempered glass: a few radial cracks extend from the center of the fracture, and accordingly the fragments are large. Because the risk of serious injuries from cuts is larger than with TSG, HSG may not be rated safety glass.

Unlike TSG, HSG is not subject to spontaneous breakage from inclusions of nickel sulfide crystals. In exceptional cases, it is possible to treat (grind) the edges of HSG, but normally this should be done before tempering.

›

**Chemical processing**

Compressive stress on the surface of glass can be produced chemically by dipping the glass into an electrolytic fluid. This process can even be used to temper very thin glass with spatially complex geometry. Chemically tempered glass has, however, only very minor significance in architecture.

**Laminated glass and laminated safety glass**

Laminated safety glass (LSG) is composed of at least two panes held together by a polyvinyl butyral (PVB) film. › Figs. 7 and 8

Such glass is often used in architecture, for overhead glass or other glass that prevents people from falling, and in the automotive industry. One essential reason for its use in such applications is that LSG holds

\\Note:
Note: The bending strength of HSG is 70 N/mm$^2$, that of TSG 120 N/mm$^2$ and that of normal glass 40 N/mm$^2$. Thermal shock resistance is 40 K for normal glass, 100 K for HSG, and 150 K for TSG. The following rule of thumb is used to estimate resistance to shock, pressure, and temperature: Factor 1 for normal glass, Factor 2 for HSG, and Factor 3 for TSG.

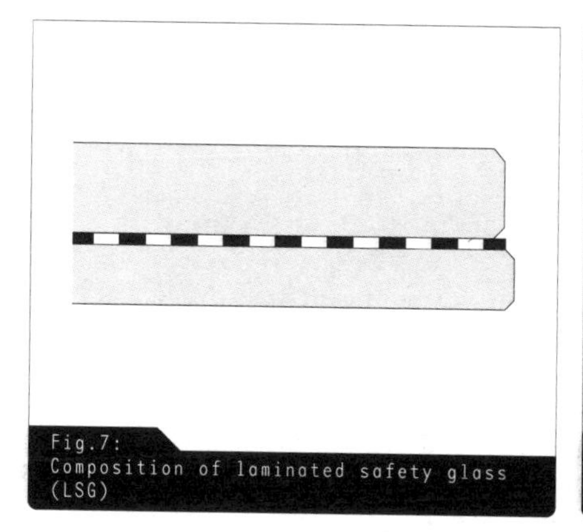

Fig.7:
Composition of laminated safety glass (LSG)

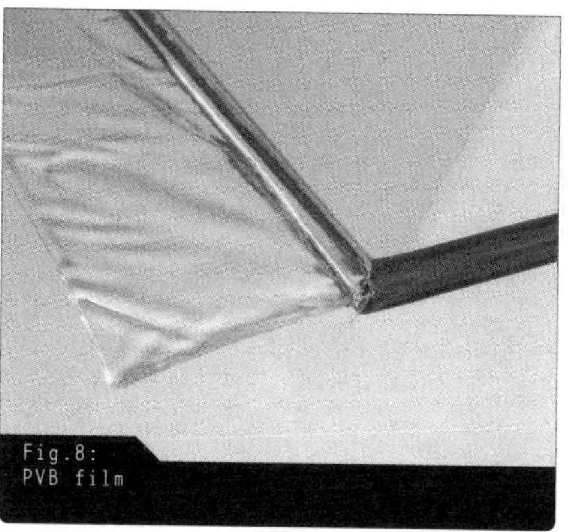

Fig.8:
PVB film

splinters together. When the pane breaks, the splinters of glass generally remain stuck to the transparent PVB film. It is difficult to pierce the film, which considerably reduces the risk of injury.

The viscoelastic PVB film is characterized by good adherence to glass, high tear strength, and high transparency and resistance to light. The nominal thickness of PVB film is 0.38 mm but it can be made many times that thickness if requested. Hence nominal thickness of 0.38 mm, 0.76 mm, and 1.52 mm are available. LSG is produced in three steps. First, a preliminary lamination between the glass and the film is produced in a clean room. Then the glass and film pass through the preliminary laminating oven in which the temperature and the pressure of the rollers increase the adherence of the glass and the film. The final laminate is created in an autoclave under high pressure and intense heat. Only then does the previously cloudy film obtain its high transparency.

Multilayer glass with interlayers made of materials other than PVB is generally referred to as laminated glass. Such glass does not usually meet the requirements for safety glass, and hence they may not be identified as such without special proof. Photovoltaic modules can be integrated into laminated glass, in which case ethylene-vinyl acetate (EVA) film is used.

**Multilayer insulating glass**

To improve heat insulation of the windows of buildings, insulating glass is now used almost exclusively. The linear lamination of two or more sheets of glass along the edge is called multilayer insulating glass (MIG). Dehumidified air or inert gas is hermetically sealed in the air space between the pane.

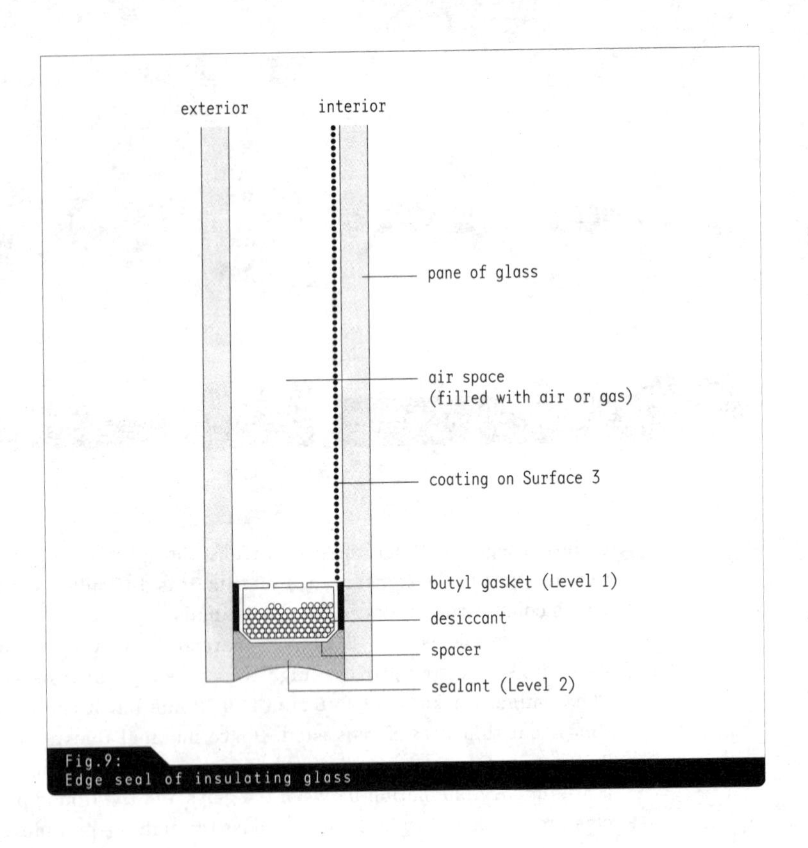


Fig.9:
Edge seal of insulating glass

Edge seal

The edge seal running around the pane consists of a spacer of aluminum, stainless steel, or plastic and sealants. › Fig. 9

The spacer is filled with a hygroscopic dehumidifier that absorbs the residual humidity in the gas or air. This helps keep condensation from forming in the air space. The seal is provided by two levels of sealant. The first level consists of a butyl sealant that also glues the spacer to the glass; the second level consists of a permanently elastic sealant such as polysulfide, polyurethane, or silicon.

These days, the inert gases argon or krypton, and more rarely xenon, are usually used to fill the air space between the panes, since inert gas improves the insulation of the glass as compared to dehumidified air. It is, however, the coating of the surface of the glass that markedly improves the heat insulation that insulating glass provides.

The effect of insulating glass

Because the air in the air space of insulating glass is hermetically sealed, differences in pressure between the trapped gas and the atmosphere can cause the panes to bend inward or outward. › Fig. 10

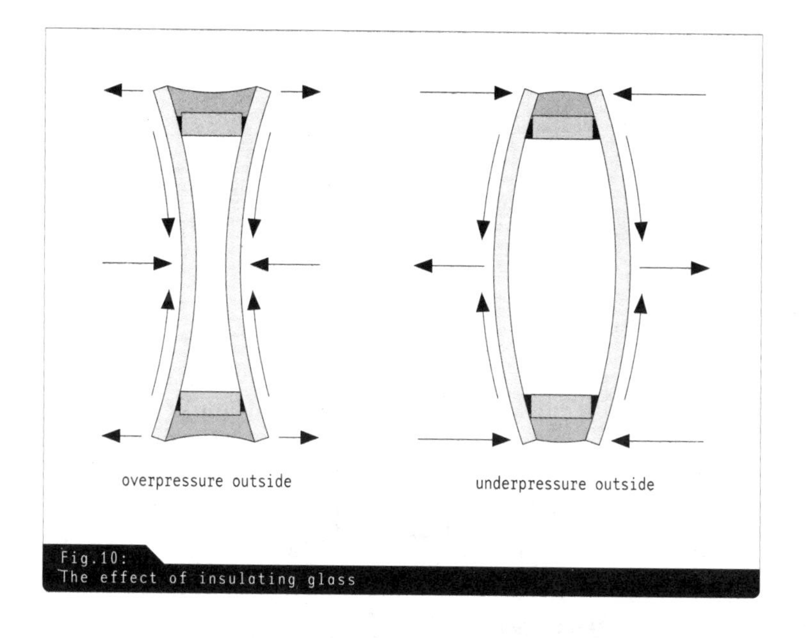


Fig.10:
The effect of insulating glass

Consequently, reflections on the pane will be distorted. The differences in pressure result in additional stress on the panes, known as climatic stress. This places strain on the glass and especially the edge seal. In the case of small or very narrow formats in particular, this effect of insulating glass can also cause overstressing and premature failure of the edge seal.

### Coating the surface

Coating the surface of the glass can considerably influence the optical and physical properties of the glass, depending on the requirements. In general, metals and metal oxides are used as coating materials. Today, types of glass can be produced that provide insulation or solar control, reflect or do not reflect, provide colors, or repel grime. The thin layers do not affect the structural properties of the glass, but the coatings are often not resistant to environmental influences (corrosion) or mechanical influences (scratching). Thus, many layers, especially those effective as insulation or solar control, can only be applied on the surface of the pane that faces the air space (that is, Surface 2 or Surface 3 in fig. 11). › Fig. 11

The coating is applied either during the manufacture of float glass ("online," as it were) or after manufacture is complete (that is, "offline"). In the online method, liquid metal oxides are applied to the surface of the glass while it is still hot, which binds them to the glass (pyrolysis). The

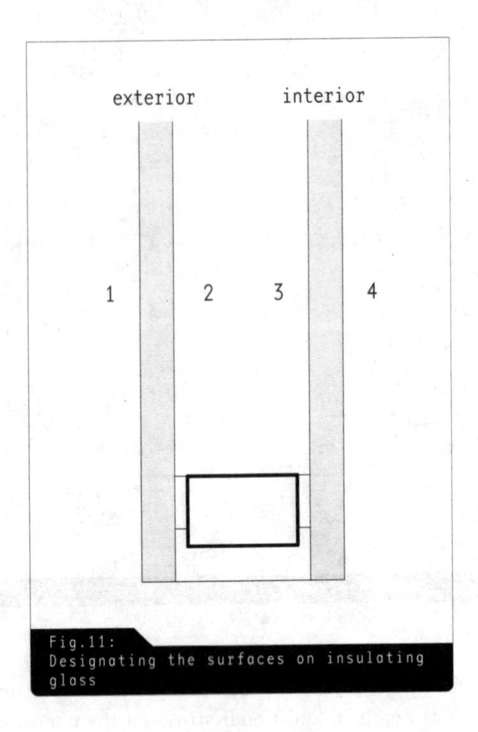

Fig.11:
Designating the surfaces on insulating glass

result is a very resistant layer ("hard coating") that is also suited to the outer surface of the pane (Surface 1 in fig. 11).

Nowadays, most glass for insulation or protection against the sun is produced by the modern cathode ray method (magnetron sputtering), since this method can apply multiple layers that are extreme thin (with thicknesses measured in nanometers). Glass can also be coated using the sol-gel process, in which the glass is dipped multiple times in a chemical solution. After each dipping, the layer absorbed by the surface of the glass is fired.

### Designing surfaces

In addition to surface coatings, there are a variety of other treatments for decorating glass surfaces. One popular way of producing custom designs in color is printing on all or part of the surface of the glass. In addition to printing, etching and sandblasting the glass can produce a translucent surface. › **Chapter Design glass**

## THERMAL INSULATING GLASS

The use of insulating glass has become absolutely a matter of course for buildings in cold and moderate climate zones because of energy conservation requirements. In the 1980s glazing was still responsible for high annual heating costs. Today even buildings with generous glazing can have very low or even zero fossil energy usage.

Coefficient of thermal conductivity

The unit of measure to indicate the heat loss of building materials, and hence windows and glazing, is the coefficient of thermal conductivity (U-value) in accordance with EN 673. The coefficient of thermal conductivity indicates the flow of warmth passes through 1 $m^2$ of a building material with a temperature differential between room and outside air temperature of 1 K (unit = $W/m^2K$). The crucial factors for the insulating properties of a window are, however, the structure of the glass and that of the frame. For that reason, a distinction is made between the $U_g$-value (g = glazing) of the glazing and the $U_w$-value (w = window) of the entire window. Because the transfer of heat is generally higher at the edge of the window than in its center, the $U_w$-value is also higher, and hence poorer, than the $U_g$-value. $U_w$ is calculated as follows:

$$U_w = \frac{U_g \times A_g + U_f \times A_f + \varphi \times L_g}{A_w} \ (W/m^2K).$$

$U_w$ = Coefficient of thermal conductivity of window
$U_g$ = Coefficient of thermal conductivity at center of pane
$A_g$ = Area of glass
$U_f$ = Coefficient of thermal conductivity of window frame
$A_f$ = Area of frame
$\varphi$ = Linear coefficient of thermal conductivity of edge of glass
$L_g$ = Length of edge of glass
$A_w$ = Area of entire window

The $U_g$-value of a pane of insulating glass composed of two panes with no coating but air space between them will be approx. 2.8–3.0 $W/m^2K$. If filled with inert gas and coated (generally on Surface 3), by contrast, the $U_g$-value will be approx. 1.1 $W/m^2K$ if filled with argon, or 1.0 $W/m^2K$ if filled with krypton. Three-pane insulated glass has a value of about 0.6 $W/m^2K$ if filled with argon or 0.5 $W/m^2K$ if filled with krypton. › Tab. 3

With insulating glass, one-third of the thermal conductivity results from convection and conduction and as much as two-thirds from radiation. Energy transportation in a gaseous medium is called "convection." Because of the temperature differential between the two panes, the gas

Tab.3:
Characteristic values of various types of insulating glass

| Structure (mm) / Composition | $U_g$-value (W/m²K) | G-value (%) | Lt-value (%) |
|---|---|---|---|
| Two-pane insulating glass 4/16/4 argon | 1.1 | 63 | 80 |
| Two-pane insulating glass 4/10/4 krypton | 1.0 | 60 | 80 |
| Three-pane insulating glass 4/14/4/14/4 argon | 0.6 | 50 | 71 |
| Three-pane insulating glass 4/12/4/12/4 krypton | 0.5 | 55 | 72 |

$U_g$-value = Coefficient of thermal conductivity of glass
G-value = Solar heat gain coefficient
Lt-value = Light transmittance
The values given here are specific to different manufacturers and hence are not universally valid.

in the air space begins to move and thus transports the warmth from the warm pane to the cold one. Heat conduction is the transportation of energy from a solid body—in this case, the energy flow through the glass—and the edge composite. The thermal radiation of glass panes includes the direct exchange of radiation between the warm glass surface and the cold one. › Fig. 12

Thermal protection coating

The thermal protection coating serves to reduce energy loss from thermal radiation, which is why this layer is also referred to as a Low-E layer (from Low-Emissivity). Silver has become established as the most common coating, as it has not only extremely low emissivity but also high great color neutrality and light transmission. Thermal insulating glass is therefore almost indistinguishable from uncoated insulating glass with the naked eye. › Fig. 13

Condensation

Although insulating glass provides significantly better thermal insulation, condensation can still form on the edge of the surface of the glass on the interior side when external temperatures are low. Over time, the condensation can damage the seal, or the glazing bead in the case of wooden windows. The cause of condensation is the higher thermal conductivity on the edge, since the spacer, which is usually made of metal, is a thermal bridge. Using stainless steel or plastic in place of aluminum for the spacer reduces the linear coefficient of thermal

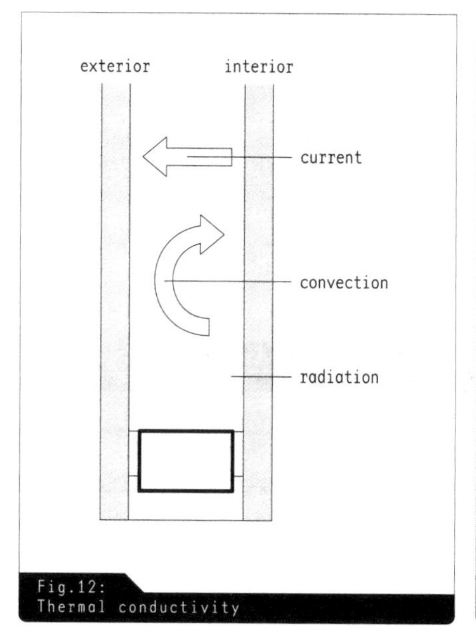

Fig.12:
Thermal conductivity

Fig.13:
Cigarette lighter test

conductivity (f) at the edge of the glass and thus reduces condensation and improves the $U_W$-value of a window by several percent, depending on its dimensions.

Solar heat gain coefficient

One of the essential characteristic values of insulating glass, along with U-value, is solar heat gain coefficient (G-value): G-value (according to EN 410) indicates the total energy admitted of the solar radiation that strikes the glass. The G-value is the sum of the primary transmittance of solar radiation and the fraction of solar radiation absorbed by the glass (secondary transmittance), in the form of thermal radiation and convection. In passive solar buildings, a high G-value is often desirable in order

\\Tip:
The position of the coating of insulating glass can be determined subsequent to installation by means of the cigarette lighter test. The flame is reflected twice by each pane. The mirror image of the coated surface differs in color from the others (see Fig. 13).

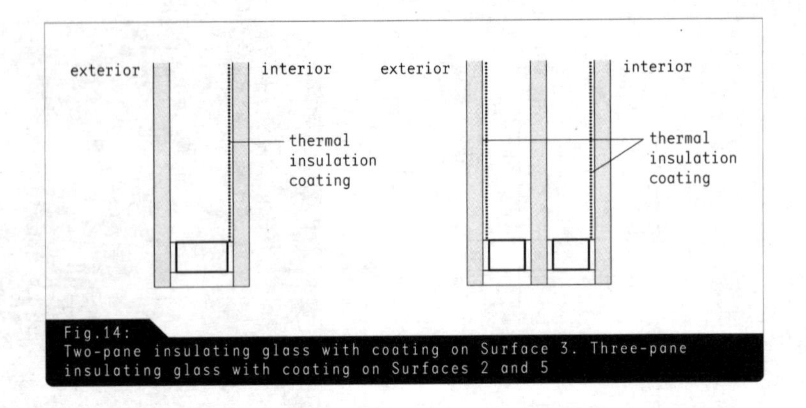


Fig.14:
Two-pane insulating glass with coating on Surface 3. Three-pane insulating glass with coating on Surfaces 2 and 5

to optimize passive solar energy gain. In other types of buildings, such as office buildings with a large percentage of glazing, too much absorption of solar energy poses a risk of overheating the interior. › Chapter Solar control glass

The G-value of thermal insulating glass lies approximately between 0.6 and 0.65. Three-pane insulating glass has a somewhat lower G-value. › Fig. 14

›

### SOLAR CONTROL GLASS

Greenhouse effect

As a transparent building material, glass is permeable to shortwave solar radiation (wavelengths from 300 to 3000 nm) but impermeable to longwave thermal radiation (>3000 nm). Much of the solar radiation that enters a room is absorbed by the surfaces on which it shines, transformed into heat, and radiated again in the form of longwave thermal radiation. This can no longer be transmitted back outside by the glass, so the room grows continuously warmer. This greenhouse effect means that glazed rooms can overheat even when the outside air temperature is low. › Fig. 15

\\Note:
Altering the position of the layer on insulating glass changes the G-value as well. With two-pane insulating glass, the G-value is decreased if the layer is on Surface 2 rather than Surface 3. With three-pane insulating glass, the G-value is higher if Surfaces 3 and 5 are chosen for the layer (rather than 2 and 5). But these measures do not affect U-value.

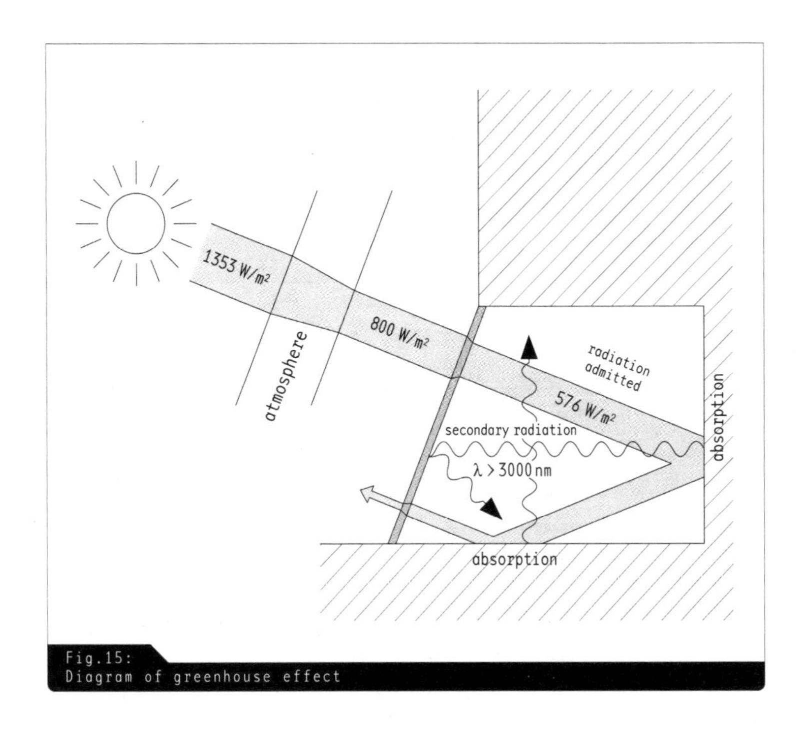


Fig.15:
Diagram of greenhouse effect

Solar control glass prevents much of the radiant energy from entering the room, first, by absorption, and second, by reflection of the radiation that strikes it. Solar control glass was often made of body-tinted glass; it absorbed part of the radiation, which unfortunately included visible light. The earliest types of coated solar control glass had the disadvantage that they reflected much of the visible light.

Selectivity

Modern solar control glass has a selective coating; that is to say, it is transparent to the visible light spectrum but reflects or absorbs long-wave infrared radiation. Today its G-value ranges from about 20 to 50%.

In order to obtain high-quality natural lighting in the interior, it is also important to look for high selectivity (S) when choosing solar control glass. Selectivity is the ratio of light transmission to G-value. For example, if the glazing has a G-value of 40% and allows 76% of visible light through, its selectivity is the quotient of 76:40, or 1.9. The theoretical limit is a value of 2.0.

Color rendering

It is not just the quantity of light that is crucial to the quality of natural light in an interior but also the color rendering. The color rendering index ($R_a$) should be at least 90%; this is the measure of the rendering of the colors of natural light, measured at the surfaces in the room where

Tab.4:
Characteristic values of various types of insulating glass

| Structure (in mm) / composition | $U_g$-value ($W/m^2k$) | G-value (%) | LT-value (%) | RA |
|---|---|---|---|---|
| Single-pane glazing 6 mm | | | | |
| Clear | 5.7 | 56 | 45 | - |
| Green | 5.7 | 45 | 53 | - |
| Two-pane insulating glass<br>Argon<br>Nominal value 68/34 | 1.1 | 36 | 66 | – |
| Two-pane insulating glass<br>Argon<br>Nominal value 40/21 | 1.1 | 22 | 40 | 88 |
| Two-pane insulating glass 6/16/4<br>Argon<br>Nominal value blue 50/27 | 1.1 | 29 | 50 | 95 |

$U_g$-value = Coefficient of thermal conductivity of glass
G-value = Solar heat gain coefficient
LT-value = Light transmittance
$R_a$ = Color rendering index
The values given here are specific to different manufacturers and hence are not universally valid.

daylight is reflected. The color rendering index ($R_a$) of glazing can be as high as 99%. › Tab. 4

› Diminution factor $F_c$

In many cases, solar control is also improved by sun shading inside or outside (slats, venetian blinds, awnings, etc.). The solar heat gain

\\ Tip:
Solar control glass, including neutral solar control glass, differs according to composition in the degree of reflection and color nuances when viewed from outside (e.g. blue, green, green, or silver). Even when its characteristic values are known, the composition should be sampled prior to installation. This is particularly important when single panes have to be replaced.

Tab.5:
Diminution factor $F_c$

| Line | Type of shading | $F_c$ |
|---|---|---|
| 1 | No shading | 1.0 |
| 2 | Interior or in air space between panes | |
| 2.1 | White or reflective surface with minimal transparency | 0.75 |
| 2.2 | Bright colors or minimal transparency | 0.8 |
| 2.3 | Dark colors or high transparency | 0.9 |
| 3 | Exterior | |
| 3.1 | Rotating slats, ventilated behind | 0.25 |
| 3.2 | Venetian blinds and fabrics with minimal transparency, ventilated behind | 0.25 |
| 3.3 | Venetian blinds in general | 0.4 |
| 3.4 | Roller shutters, window shutters | 0.3 |
| 3.5 | Canopies, loggias, free-standing slots | 0.5 |
| 3.6 | Awnings, ventilated above and on the sides | 0.4 |
| 3.7 | Awnings in general | 0.5 |

coefficient of the glazing with sun shading is called $g_{total}$. It is the product of the g-value of the glazing and a diminution factor: › Tab. 5

$$F_c\ (g_{total} = F_c \times g)$$

The G-value of glazing can also be reduced by ceramic printing on (part of) the outer pane of glass. The greater the density or coverage of the printing, the lower the G-value. › Chapter Design glass, Colored glass

## SOUNDPROOF GLASS

Architectural projects often have to meet a minimum standard for soundproofing. The degree of soundproofing required depends on the needs of the specific use. A general distinction is made between airborne noise and impact noise. Impact noise is transmitted by walking or pounding directly on parts of the building. Noises transmitted through the air, such as talking or traffic noise, are called airborne noise.

Sound pressure level, measured in decibels (dB), is the unit by which noise is measured. The resultant sound reduction (in dB) of a facade depends essentially on the soundproofing quality of its windows, that is, on the construction of the glass and the frame. Normal insulating glass already has significantly higher soundproofing than ordinary glass. The following measures can improve the sound reduction index of two-pane insulating glass even more:

- Increasing the quantity of glass and asymmetrical structure (variations in thickness of glass)
- Increasing the air space between panes
- Using laminated safety glass
- Using laminated glass or laminated safety glass (LSG) with special interlayers such as films or cast resin

Soundproofing films meet the same safety requirements (tensile strength and splinter resistance) as traditional PVB films, so that soundproofing glass with cast resin as an interlayer is only rarely used in buildings. Cast resin has only limited tensile strength, and there is a risk that over time it can it can detach and run, becoming visible on the edge of the laminated glass. > Fig. 16

### FIRE-RESISTANT GLASS

If a wall is supposed to be flame-retardant or fire-resistant, fire-resistant glass is used in the wall or facade openings. It is important to pay attention to the required fire-resistance rating of the glass according to EN 13501-1. > Tab. 6

The figures indicated in the table show the certified fire-resistance rating of glass. This is the period of time in which glass at least slows the penetration of fumes from fire, indicated in minutes (30 or 90 minutes). E glazing stops only the penetration of fumes, while EI glazing also prevents the penetration of the high thermal radiation from the fire. E glazing can

\\ Note:
The weighted sound reduction index ($R_w$) is the sound reduction index of a unit for insulating glass that has been measured by an authority recognized in the building code and authenticated with a certificate. The value is calculated in a laboratory and does not take into account the transmittance of sound via adjoining parts of the building. The weighted sound reduction index ($R'_w$), by contrast, also takes into account the adjoining parts. This value is thus somewhat smaller than the $R_w$.

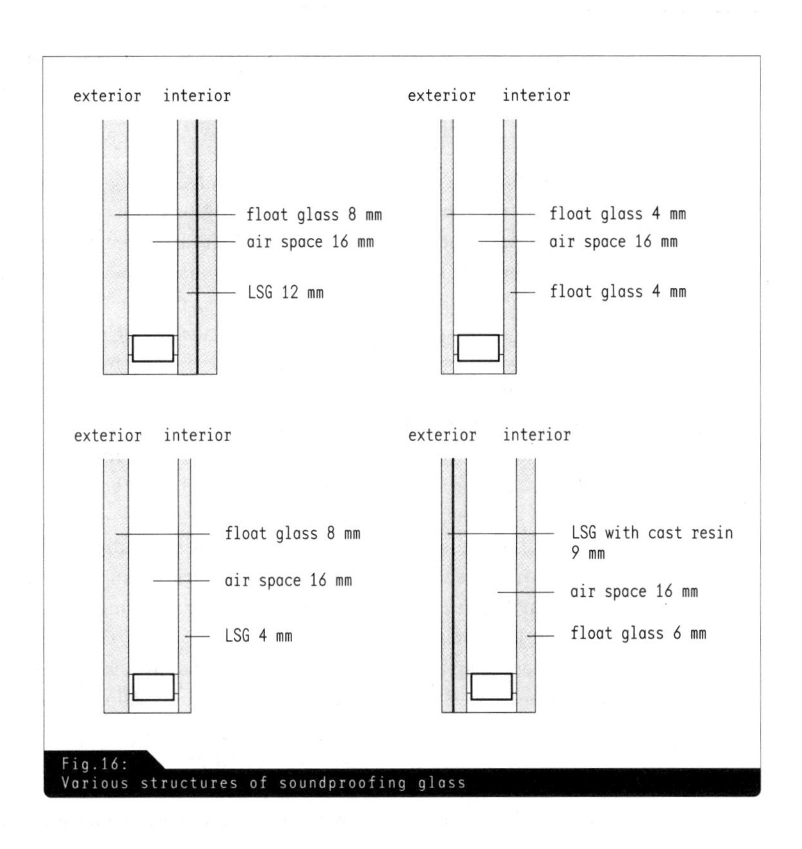

Fig.16:
Various structures of soundproofing glass

Tab.6:
Classification of fire-resistant glass

| Building code designation | Fire-resistance rating according to EN 13501-1 |
| --- | --- |
| Flame-retardant | EI 30<br>E 30 |
| Fire-resistant | EI 90<br>E 90 |

therefore only be used where there is a sufficiently safe distance from people in case of fire—for example, in skylights more than 1.80 m above the floor or in walls that do not adjoin escape routes. Whereas E glazing is monolithic in construction, EI glazing has a multilayer structure composed of normal plate glass and special interlayers that foam up at high temperatures and thus prevent the penetration of fire and heat for the required duration.

### SPECIAL TYPES OF SECURITY GLAZING

The spectrum of requirements for glass today also includes protection against vandalism, theft, and violent crime. European standards distinguish between special types of glazing based on the nature and force of the violence against which they provide protection.

Impact resistance

Impact-resistant glazing offers protection against stones or smaller projectiles. Such glazing is tested using a falling ball experiment. A steel ball weighing approx. 4 kg is dropped onto the glass three times in a row from a specified height but should not penetrate it.

Burglary resistance

Burglary-resistant glazing prevents an ax from cutting an opening of 40×40 cm within a specified period. The test is conducted using a long-handled ax attached to a machine.

Bullet resistance

Bullet-resistant glass, commonly called bulletproof glass, is produced for protection against various firearms, ranging from shotguns to rifles. Such glazing is tested in a laboratory shooting range using common types of firearms.

Blast resistance

Blast-resistant glazing protects against attack from outside with explosives. It is tested using an artificially produced pressure wave perpendicular to the glass.

All special security glazing is composed of several layers, ranging from traditional laminated safety glass (LSG) to multilayered structures that consist of glass, tempered glass, films, or even plastic coatings. The focus, however, is always on protecting the people or objects behind the glass. The glass itself is usually damaged when subjected to severe violence and has to be replaced. Security glass can also be fitted with a conductive wire; its current is interrupted when the glass is broken, triggering an alarm. These conductive wires are either thin silver wires inserted into the LSG or conductive enamel printed on one corner of the pane of TSG. Because TSG always breaks into small pieces, the circuit is sure to be broken.

\\ Note:
To ensure fire safety, all fire-resistant glass must be framed in appropriate constructions in combination with certified materials for connecting and sealing.

### INSULATING GLASS WITH INTEGRATED ELEMENTS

One current trend is to integrate functional elements into the air space of insulating glass, for example, slats for shade or metal weave. The advantage of this is that the elements described below are protected from damage by wind and weather and from grime.

Adjustable systems

Adjustable sunshade slats are integrated venetian blinds made of concave or convex slats with matte, reflective, or perforated surfaces. They can be regulated to provide protection against sun and glare at workplaces with computer monitors.

Adjustable sunshades can also be integrated into the air space. Such blinds are made of textiles or perforated plastic film.

Fixed systems

Fixed sunshade slats are highly reflective, horizontal slats with a special cross section preinstalled in the air space so that most of the direct sunlight is reflected, while diffuse daylight is directed into the room. › Fig. 17

Sunshade grids function like such fixed sunshade slats, except that the slats are arranged vertically and horizontally. Such complex grids, coated with the purest aluminum for reflectivity, were developed especially for use in glass roofs with little slope or for domed roof lights. › Fig. 18

Integrated prism plates or acrylic glass profiles are also used to reflect direct sunlight back out while directing diffuse daylight into the room.

Furthermore, a wide variety of metal weave inserts are available. They not only provide solar control but also emphasize the design. › Fig. 19

Fig. 17: Fixed sunshade slats

Fig.18:
Sunshade grid

Fig.19:
Metal weave insert

Filigree wooden inserts consisting of wood bars of rectangular cross section can be employed in insulating glass instead of metal weave.
› Fig. 20

Finally, mention should be made of other integrated elements that merely scatter light and hence, if thick enough, are also suitable as thermal insulation. Light-scattering capillary plates in insulating glass are made of tiny, UV-stable, transparent polycarbonate tubes that scatter daylight evenly. They are sometimes used for natural lighting in buildings

Fig.20:
Wooden insert

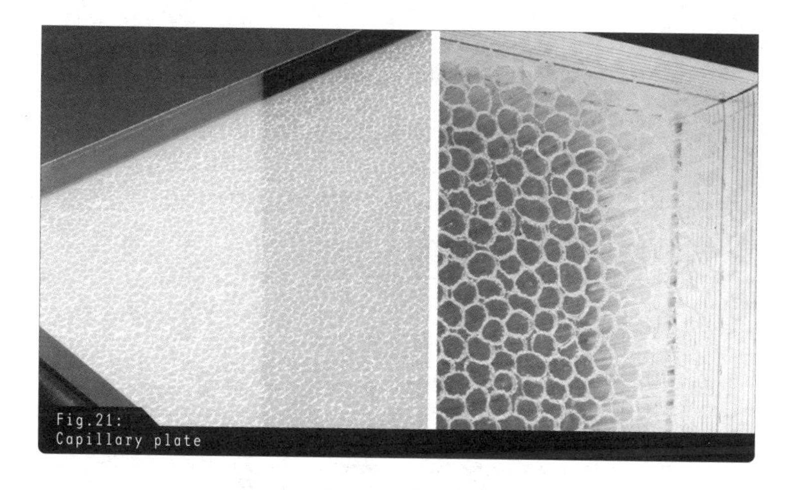

Fig.21:
Capillary plate

where direct lighting is not desirable, for example, in museums, studios, sporting arenas, and so on. › Fig. 21

Transparent thermal insulation

The term transparent thermal insulation is somewhat misleading, since this kind of thermal insulation is usually not transparent at all but merely diffusely translucent. Transparent thermal insulation is often integrated directly in front of a solid exterior wall to provide it with additional winter heating from solar radiation, the energy from which is then transferred to the interior over time. Transparent thermal insulation can

Fig.22:
Photovoltaic element

also be used as infill on the facade in order to profit from additional use of daylight. Materials such as plastic capillary plates, glass tubes with thin walls, or aerogels (foamed granules) are placed between two panes of glass. There should be some means of shading the transparent thermal insulation, however, to avoid overheating the interior on warmer days.

\\ Note:
One disadvantage of integrated elements is that they increase the thermal absorption of the air space between the panes, which can cause premature damage to the elements integrated in the glass. It can be expensive to replace the elements after only a few years. It is also important to ensure that the replacements panes look exactly the same. In integrated venetian blinds, there is also a risk that the double-pane effect (inward curving from climatic stress) on the panes cause the slats to jamb and cease to operate. The bending of the panes can be reduced by using thicker glass.

Fig. 23: Electro-optic functional layer

## SPECIAL FUNCTIONAL LAYERS

Its transparency, strength, and resistance to weather make glass an ideal base for many kinds of functional layers.

Photovoltaic

One typical example of special functional layers is photovoltaic elements, which are used in facades or on roofs to transform sunlight into electrical energy. The elements are composed of silicon cells embedded between two panes for protection, or of "thin-film cells," which are produced by coating glass directly. › Fig. 22

Switchable layers

In addition, there are functional layers that, thanks to glass's penetration by light or radiation, can adjust to current lighting or climatic conditions.

Thermotropic functional layers change how much light will penetrate the glazing according to the temperature. They can be set up, for example, so that a pane that seems transparent at room temperature looks white at higher temperatures and hence reflects much of the light diffusely.

Electro-optic functional layers can be switched between diffusely translucent and transparent at the press of a button that applies electricity. › Fig. 23

Electrochromic functional layers are switchable layers that provide continuously adjustable control of the amount of light and energy insulating glasses take in. The pane colors—e.g., turns dark blue—when the electricity is turned on, which reduces the transmission of daylight and solar thermal radiation.

Fig.24:
Custom ornamental glass

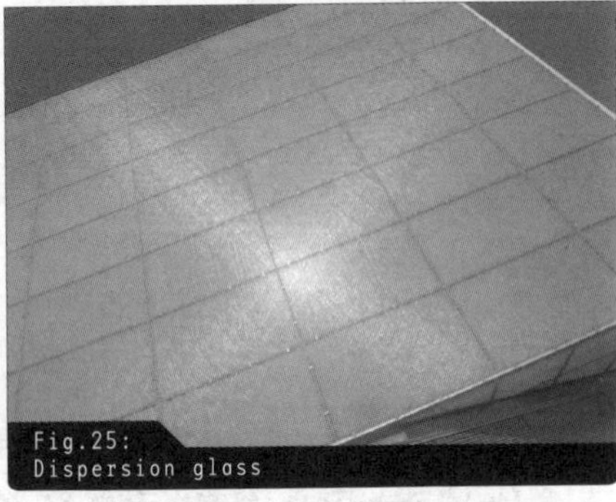
Fig.25:
Dispersion glass

## DESIGN GLASS

### ORNAMENTAL GLASS

"Ornamental glass" is the term used for cast or rolled glass given a surface structure for functional or design reasons. The variety of possible surfaces ranges from geometric patterns (squares, rectangles, lines, dots) to amorphous patterns and custom ornaments, which can be made to order if quantities are sufficient to justify the effort. › Figs. 24 and 25

Ornamental glass can be slightly or highly diffuse, and it is often used where either diffuse light is desired or clear visibility is undesirable for functional or design reasons. Ornamental glass can be made with either one or both surfaces structured. One special case is glass with a relief surface pattern (e.g. prism glass) that does not scatter daylight diffusely but reflects it or guides it in a particular direction according to the angle of incidence. › Figs. 26 and 27

\\ Note:
Most types of ornamental glass can be tempered, and many can also be used to make LSG or insulating glass. When it is used for facades, in particular, the manufacturer should be asked to verify this.

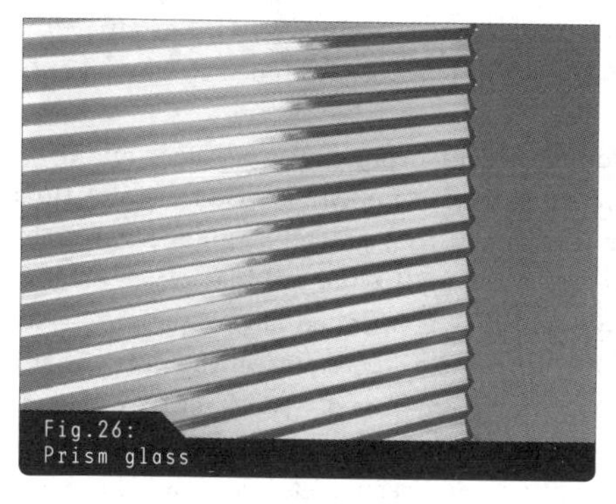
Fig.26:
Prism glass

Fig.27:
Using prism glass to direct light

### GLASS WITH A FROSTED SURFACE

Frosted glass is a popular design element in architecture, and not without reason. Frosting ensures that the play of light will be interesting, and it heightens the materiality of the glass.

Etching

Etched glass is frosted with hydrofluoric acid. The surface of the glass is damaged only slightly in the process, and hence the strength of the pane is largely preserved. The concentration of acid used today is very low, but this is compensated for by allowing it to work on the surface of the glass for a long time. The duration of application determines the amount of frosting. As the roughness of the surface increases, the transparency of the glass is reduced, since a rough surface will increase the scattering of the light that enters. Stencils can be used to etch individual designs, logos, or patterns. In consultation with the manufacturer, etched glass can be tempered or even bent.

Sandblasting

With sandblasted glass, the surfaced is roughened by sandblasting. Since the surface is damaged by the roughening, the strength of a pane is diminished. Over the course of time, changes in color can result from the roughness of the surface, for example, from grease residue after cleaning. The visual effect is similar to that of etched panes. Images or patterns can also be applied.

Frosting by screen printing is a way to treat the surface without roughening it. A translucent enamel paint is applied and burned in to make it permanent.

### COLORED GLASS

Colored glass has recently been enjoying a renaissance in architecture. As early as the Middle Ages, the light in the interiors of Gothic cathedrals was artfully altered by means of tracery windows subdivided into colors. These days there are many kinds of colored glass, with various technical and visual qualities. Depending on the product, the "dyeing process" takes place when the glass is produced or during subsequent processing.

Body-tinted glass

Body-tinted glass is produced by incorporating additives (metal oxides) directly into the glass paste. This method can be used to color float glass, sheet or window glass, and cast glass, as well as glass bricks. The range of colors for float glass is, however, limited to blue, green, bronze, and gray, while the other kinds of glass mentioned have a wider variety of options. Because of the natural iron oxide content in the glass paste, neutral float glass already has a slight green cast—an effect that is more evident when the glass is thick or frosted. White glass is the term for float glass low in iron oxide that has no green cast. Because it is highly transparent, it is also used to produce sun collectors and photovoltaic elements.

Fusing

Fusing is the process of combining glass made from paste of different colors to form a single pane. › Fig. 28 Sheets of glass of various colors and shapes are assembled into a larger sheet and then fused in a kiln at temperatures up to 1500 °C. This technique can be used for sheet glass but not for float glass.

Dichroic glass

Optical effect filters or dichroic filters are thin layers of various metal oxides of different thickness that are applied to glass using the sol-gel method (dipping in a chemical solution). The effect of color results from interference between the individual thin layers and varies according to the angle of incidence of the light. A glass facade of dichroic glass can, for example, reflect sunlight as cobalt blue or gold depending on the angle of incidence. Color effect filters are not very absorbent, that is to say, they will not heat up from solar radiation as much as dyed glass, for example.

Enameled glass

Enameled and screen-printed glass are types of color-coated glass in which a colored enamel layer is burned into the surface of the glass during the production of TSG or LSG (at temperatures over 600 °C). Enameled or

\\ Note:
Body-tinted glass will heat up from solar radiation and thus is often tempered for use in facades. By contrast, plate glass produced by fusing cannot be tempered because its seams are irregular.

\\ Note:
Dichroic glass is available in sizes up to approx. 1.70 × 3.80 m. Tempering is possible but only within strict limits. The layer is durable and scratch-resistant, but it should not be exposed to weathering.

Fig. 28:
Colored glass window of fusing glass

screen-printed glass is thus always tempered; printing without tempering is possible only with organic colors of two components, but it is not scratch-resistant. The best color quality is achieved by using white glass, which is low in iron oxide. It is important to distinguish between the three methods of applying enamel to glass continuously:

- In the rolling process, the flat pane of glass is passed through a grooved rubber roller that transfers the enamel paint to the surface of the glass.
- In the casting process, the sheet of glass is passed through a dip coater that covers the surface with paint. This method is outdated and not environmentally friendly, since, unlike the other methods, it is impossible without the use of solvents.

Screen printing

The most uniform application of color is achieved by screen printing. Here the paint is pressed through a fine-mesh screen onto the surface of the glass on the printing table. A wide variety of standard colors is available. These colors include opaque, transparent, and translucent colors as well as custom paints that can be applied either continuously

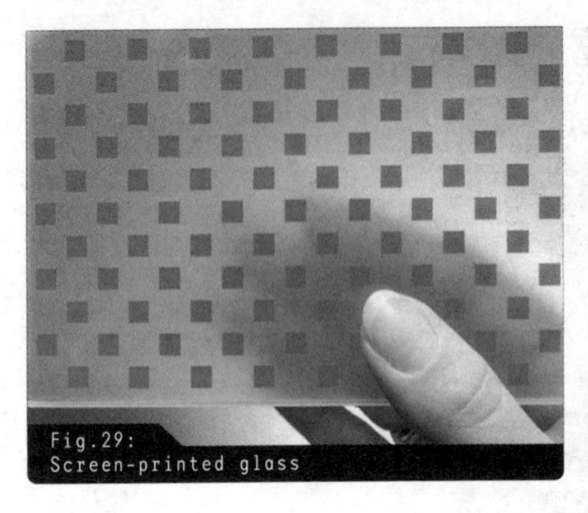
Fig.29: Screen-printed glass

Fig.30: Facade with screen-printed glass

or discontinuously. Using the available decorative patterns and stencils, it is possible to design the glass panes individually. In computer-to-screen (CTS) imaging, the screens are produced from digital designs or photographs. Multicolor printing motifs require a corresponding number of screens and passes. This means, for example, that for a four-color screen print intended to print a photograph on glass, four different screens and passes are required. › Figs. 29 and 30

Digital printing

It has recently become possible to print glass with ceramic using digital printing. The advantage is that the data of the image is sent directly to a special printer, and no expensive screens need be produced. Moreover, multiple colors can be applied at the same time. This method is particularly well suited to elaborate custom motifs.

LSG with colored film

LSG with colored interlayers is often used today instead of body-tinted glass made with colored paste. Rather than treating the glass itself, the color is introduced by means of colored films glued between two layers of glass.

A composite of glass, film, and glass behaves like a traditional laminated safety glass, since polyvinyl butyral (PVB) is used as the raw material for the extruded color films. As many as four films can be combined between two sheets of glass. Hence it is possible to generate more than a thousand transparent, translucent, and opaque colors from eleven basic colors. In addition to monochrome and patterned films, there are also films with high-resolution digital printing laminated into LSG. The opacity of the color is, however, lower than that of screen printing on glass. › Fig. 31

Holographic films

Holographic optical element (HOE) is a term for laminated glass embedded with films that have holographic grids. HOEs have an effect similar to prisms, in that they break down white light into its spectral colors. The

Fig.31:
LSG with patterned film

effect of the color depends on the angle of incidence and the viewing angle. They can produce dynamic color effects, much like the dichroic filters discussed earlier. Holographic optical films can be employed outdoors only if protected within laminated glass. They are used not just for color effects but also to direct daylight and to provide solar control. They are also suited to increasing photovoltaic electricity production by focusing sunlight onto solar cells.

\\Note:
Enameling reduces the bending strength of TSG by about 40%. Up to four passes in different colors are applied to a surface. Formats up to about 3.00 × 6.00 m are possible. Custom motifs are more expensive as the custom screens have to be produced. The printed side of the pane is scratch- and weather-resistant, but the color can change over time because of ultraviolet radiation. The printed side of insulating glass is thus usually placed on Surface 2. The color and the amount of printing change in relation to the solar heat gain coefficient of the glazing, which is why screen-printed glass is also used for solar control.

\\Note:
LSG with color film usually absorbs less warmth than body-tinted glass; nevertheless, with some dark colors it is advisable to temper the individual layers. Protected within the laminate, the colors generally remain stable when exposed to ultraviolet radiation.

# CONSTRUCTION AND ASSEMBLY

## GENERAL

Because glass is brittle, great care is required, as well as knowledge of materials, when planning and constructing glass structures. Unlike many other tough building materials such as wood or steel, glass can break immediately if it strikes a hard object. For this reason, the bearing capacity of glass components and structures is often carefully tested in a laboratory. This process includes testing their residual bearing capacity—that is, the level of stability and bearing capacity after a break in the glass has occurred. There are many standards, technical guidelines, and regulations on glass and glass constructions. In some countries such as Germany, building nonstandard structures or manufacturing components must be approved on a case-by-case basis.

## GLAZING WITH LINEAR SUPPORTS

Glass panes that need to be fastened along a continuous edge are supported linearly. In most cases involving windows, facades, glass roofs, and so on, the glass panes are supported linearly on all sides, or on all edges. They can also be supported on three sides, two sides, or one side. › Fig. 32 All-glass railings with a lower edge attached to the edge of the roof are one example of a one-sided support system › Chapter Applications, Fall-prevention glazing

Vertical and overhead glazing

A general distinction is made between vertical glazing (tilt angle < 10° from the vertical) and overhead glazing (tilt angle > 10° from the vertical). As a rule, laminated safety glass (LSG) is used for overhead glazing. The PVB film helps prevent shards of glass from falling on people when breakage does occur; LSG made from float glass or HSG also has a much higher residual bearing capacity than monolithic glass. Monolithic glass (such

\\ Note:
LSG made of TSG may possess high bending strength, but it is not permitted in some countries for overhead areas, because of its poor residual bearing performance. With insulation glass for overhead areas, the bottom glass pane is made of LSG. Wired glass is permitted for overhead areas only where the span of the main load direction is at least 0.7 m.

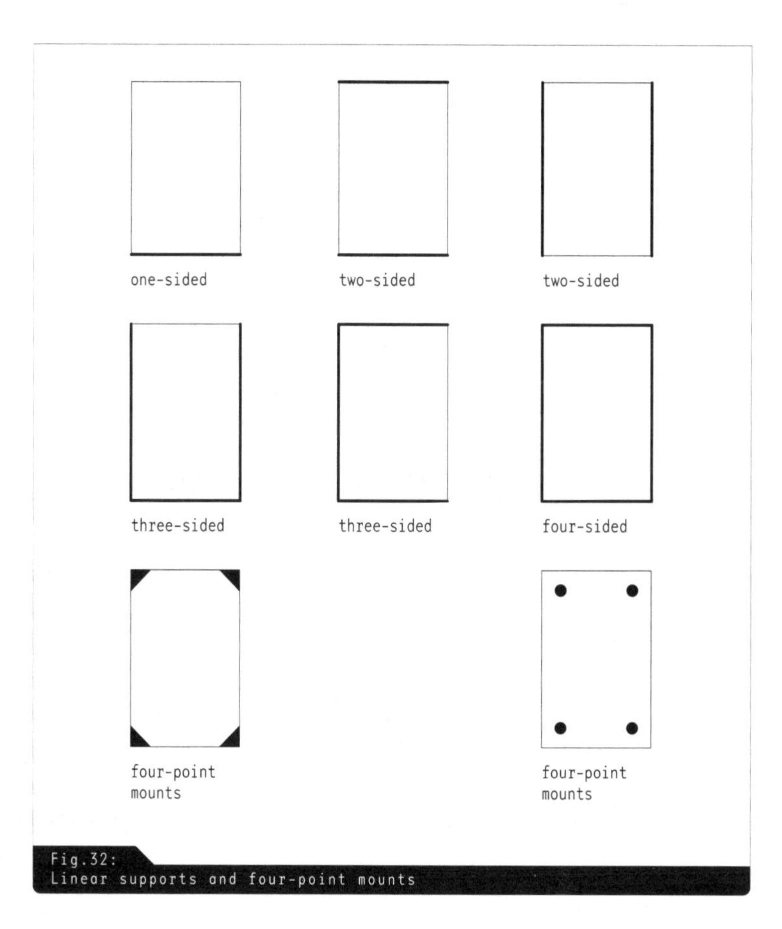

Fig.32:
Linear supports and four-point mounts

as float glass, TSG, cast glass) may be used in cases where large pieces of glass falling on public thoroughfares can be otherwise prevented, for example by installing appropriate nets with a mesh width of ≤ 40 mm.

### GLASS AND FRAMES

Glazing rabbet

The support needs to have flexible interlayers so as to ensure an even load transfer in the glass panes and to compensate for irregularities. Direct contact with hard materials such as steel or concrete should be avoided at all costs. The glass recess is the specified depth to which the glass extends into the glazing rabbet. The recess is determined by the size of the glass pane, the dimensional tolerance, and the expected deflection of the construction. › Fig. 33

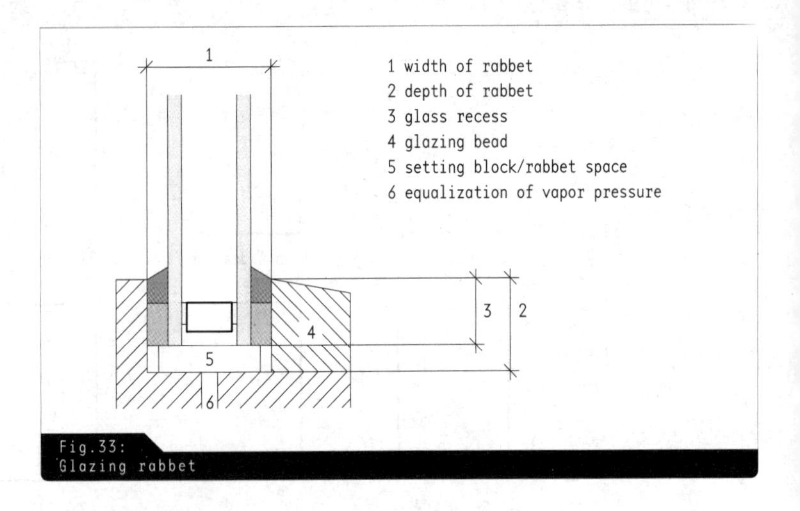

Fig.33:
Glazing rabbet

Inserting setting and location blocks

Support bolts in the glazing rabbet transfer the vertical load of the glass weight to the sash or frame. In addition, setting blocks in the frame ensure the panes cannot shift sideways. › Fig. 34

Sealing

The rabbet is now generally constructed without the use of a sealing compound, which guarantees an equalization of vapor pressure (easing of tension) inside it. Caulking is carried out using either a wet sealant applied to a sealing tape (such as silicon, acrylate, polysulfide, or polyurethane), or a dry sealant with a prefabricated gasket profile (such as synthetic rubber). Condensation that collects in the rabbet has to be able to evaporate through small vapor pressure equalization openings. › Fig. 35

It is important to consider the chemical compatibility of the different sealing compounds when planning windows or glass facade structures. There are basically five different classes of sealing compound: butyl, acrylate, polysulfide, polyurethane, and silicon. Sealing compounds vary in their chemical composition, for example in their levels of diluents, solvents, cross-linking agents, and fillers. › 

\\Note:
The wrong sealant compound can cause damage to the setting and location block material, edging tape, or sealant material. To assure the glazing structure's long-term durability, check the compatibility specifications of the manufacturer's sealant compound.

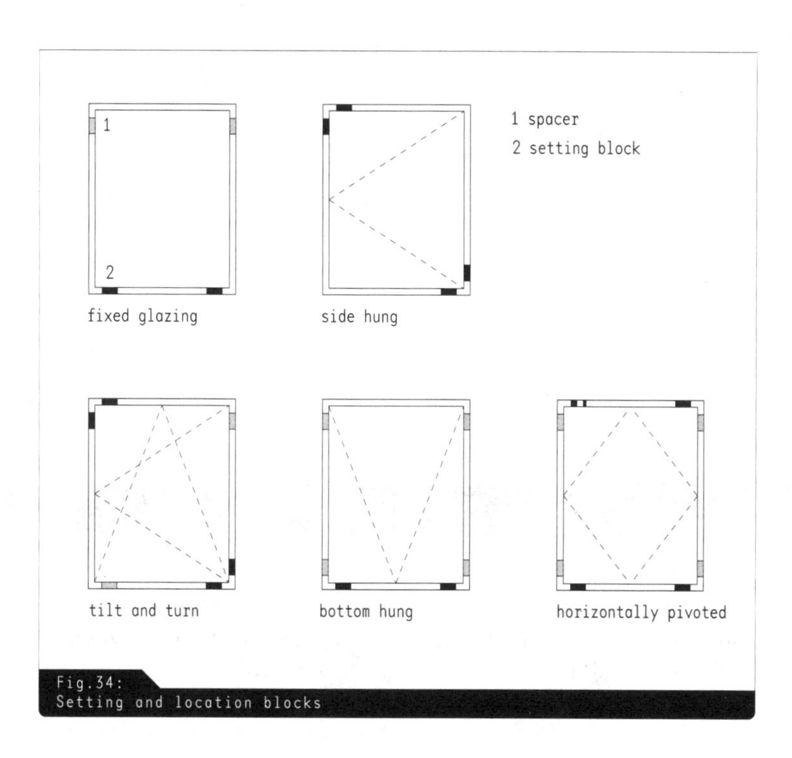

Fig.34:
Setting and location blocks

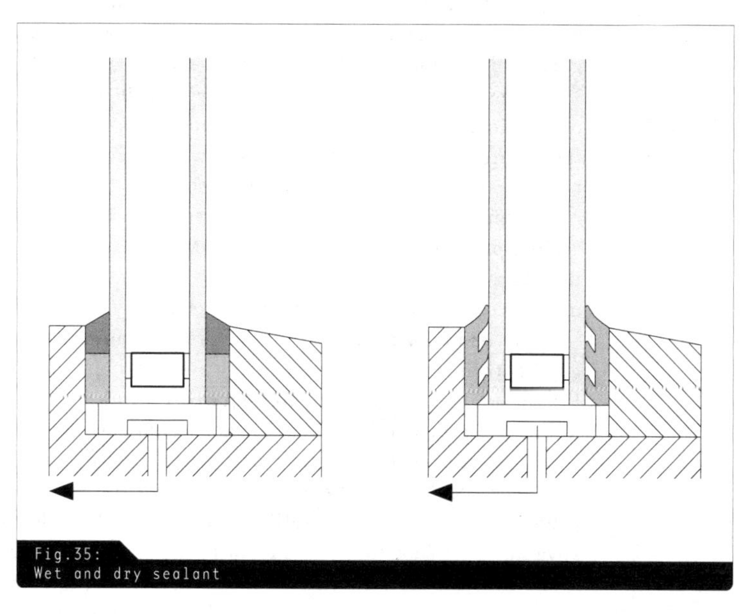
Fig.35:
Wet and dry sealant

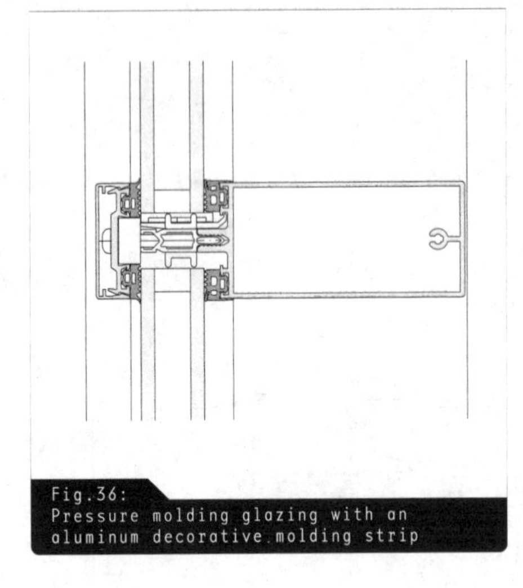

Fig.36:
Pressure molding glazing with an aluminum decorative molding strip

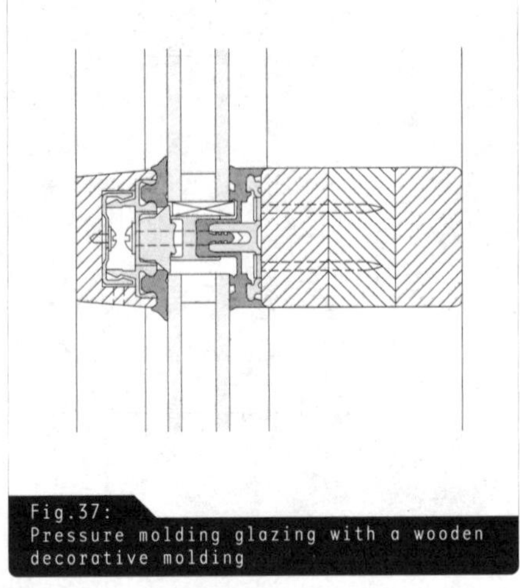

Fig.37:
Pressure molding glazing with a wooden decorative molding

Glazing beads

Glazing with linear supports can be mounted using two essentially different methods. In constructing windows and facade elements, glazing beads are the most common method of fitting glass. The glazing beads are looated on the inside of the sash. They are installed either by nails in the support structure (for example with wooden frames) or clamped (in metal or plastic frames). The contact pressure of the glazing beads secures the glass pane mounting and tightens the seal.

Pressure moldings

Pressure moldings are various beads made of aluminum, steel, wood, or plastic that are installed from the outside and press the glass against the substructure. The beads are fixed with screws that allow the contact pressure to be positioned precisely. In most systems the screws are covered by a second molding. The seal is achieved by adding a permanently elastic sealing profile made of silicon or EPDM/APTK. When insulation glass is used in windows it also requires thermal separation between the pressure molding and the support molding, for example by using a plastic insulation molding. As in a window frame rabbet, condensation can also collect in the gap between the glass joints, or rainwater from outside can leak through. For this reason, small holes are required to equalize any vapor pressure. In larger facade constructions, the horizontal and vertical joints are connected to form a communicating drainage system.

The specific material of the pressure molding can be chosen independently of the support profile's material, yet the advantages and disadvantages of each material should be weighed in relation to one other. Aluminum is more resistant to corrosion than steel, for example, and the

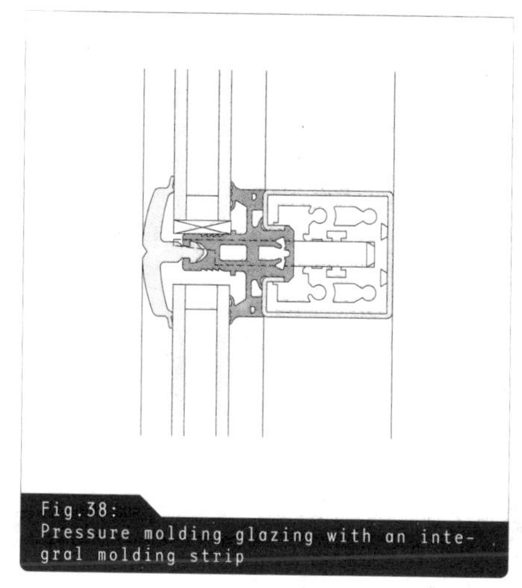

Fig. 38:
Pressure molding glazing with an integral molding strip

Fig. 39:
Facade assembled using pressure-molding glazing

extrusion press production method makes it easier to manufacture. It should nonetheless be combined with an aluminum covering to protect it from erosion.

One particular form of pressure molding is called integral molding. This is a permanently elastic, synthetic molding that combines the functions of a pressure molding and a sealing profile. › Figs. 36–39

### GLAZING WITH POINT MOUNTING SUPPORTS

One benefit of this type of assembly is its ability to deliver very delicate and transparent glass surfaces. With point mounts, the glass pane is not supported along the entire side but only at specific points. Rectangular or square glass panes are attached on at least all four corners; larger formats are additionally anchored to a substructure by point mounts. The local tension in the glass can be very great; tempered glass (TSG or HSG) is therefore generally recommended and sometimes even required. There are two different types of point mount fasteners: clamp mounts, which do not penetrate the glass, and fasteners that require a borehole in the glass.

Clamp mounts

Clamp mounts secure the pane by clamping the glass corners and/or edges. They are made of aluminum or stainless steel and are available in a range of square to round shapes. › Figs. 40–44

During the planning and constructing phase, it is important to note that any direct contact between metal and glass should be avoided by using elastic intermediary layers. Strain due to canting or too much contact pressure from the fixtures can cause the glass to break and should be

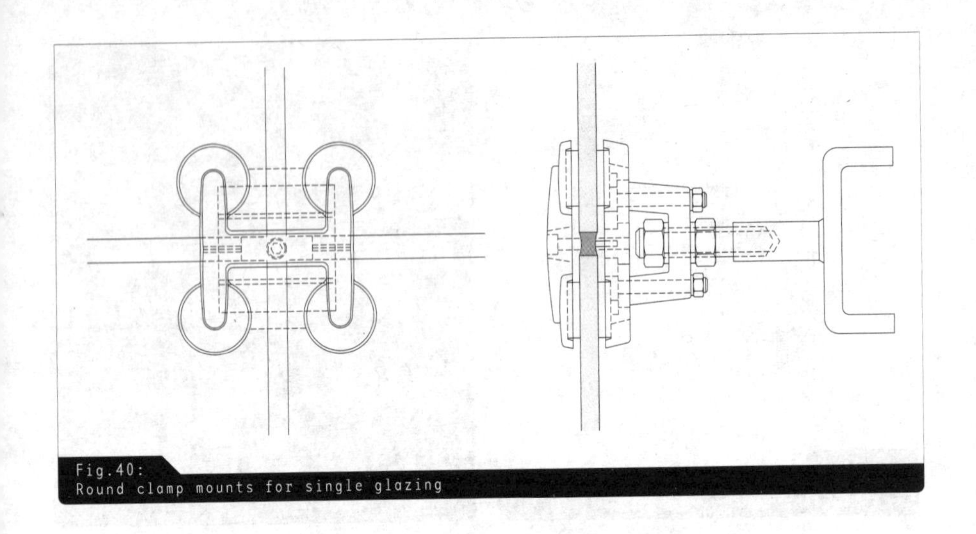

Fig.40:
Round clamp mounts for single glazing

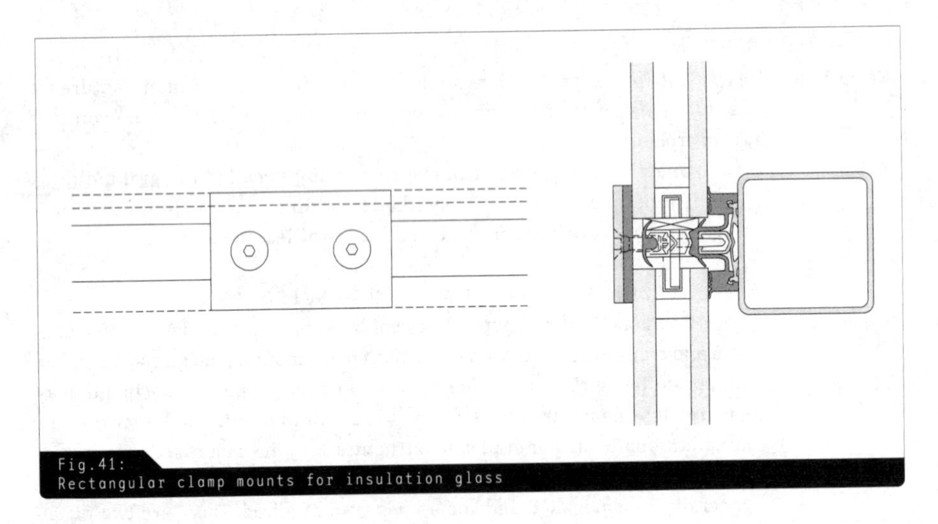

Fig.41:
Rectangular clamp mounts for insulation glass

avoided at all costs. Clamp mounts are also often custom made for building projects. According to the construction of the clamp mount, glass panels can either be flush-mounted or scaled. The clamped area covering the glass should be no smaller than 1000 mm$^2$ (with at least a 25 mm glazing rabbet recess) and is determined by the amount of glass tension expected. Since this method does not call for boreholes to be made in the glass, the clamp mounts are bolted together in the joints between the glass panes.

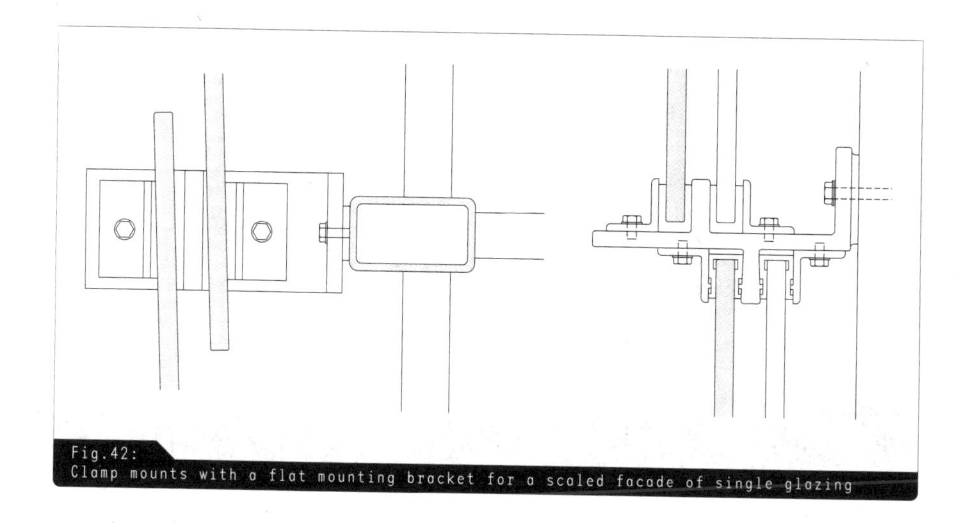
Fig.42:
Clamp mounts with a flat mounting bracket for a scaled facade of single glazing

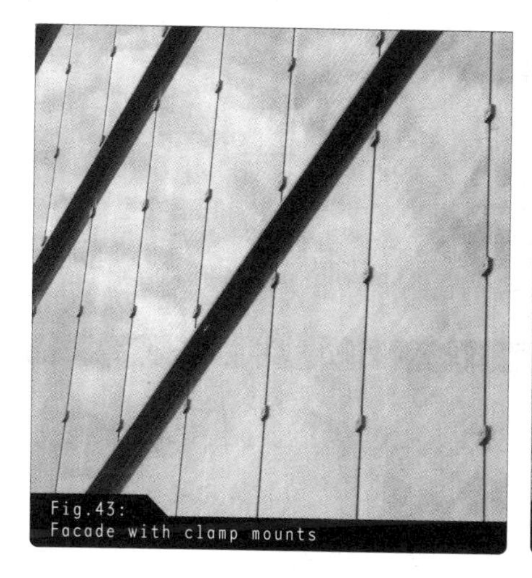
Fig.43:
Facade with clamp mounts

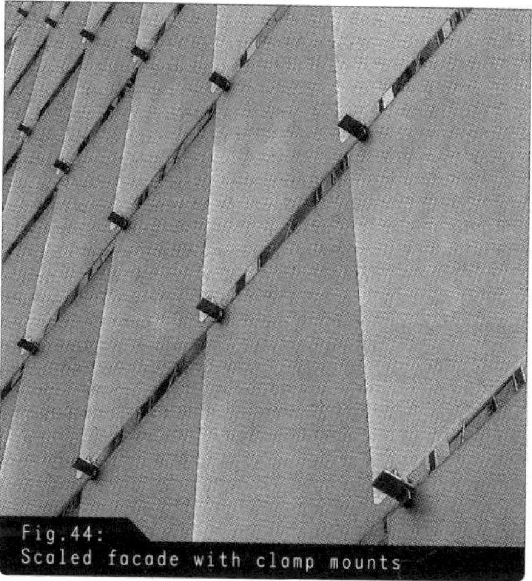
Fig.44:
Scaled facade with clamp mounts

Point mount glazing clamps can also be combined with a linear support. In this method the glass pane rests on a continuous supporting molding and is pressed by a clamp plate against the molding at particular intervals.

Point mounts in the borehole

Point mounts in the borehole secure the glass pane from within the glass surface. The boreholes required for this glazing system make it more labor-intensive than the clamp mount method. Point mounts are available

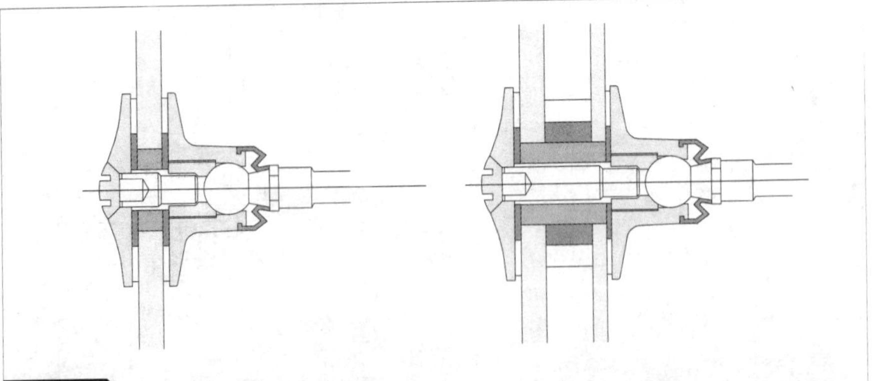
Fig.45:
Oval head point mount for single glazing and insulation glass-pin joint outside the disk level

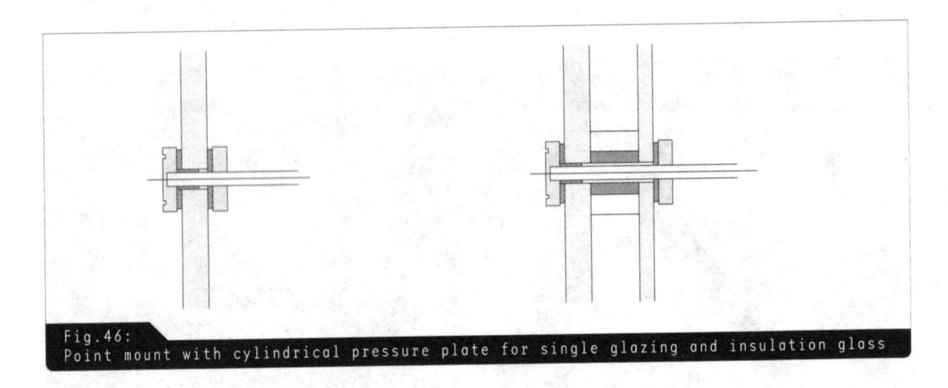
Fig.46:
Point mount with cylindrical pressure plate for single glazing and insulation glass

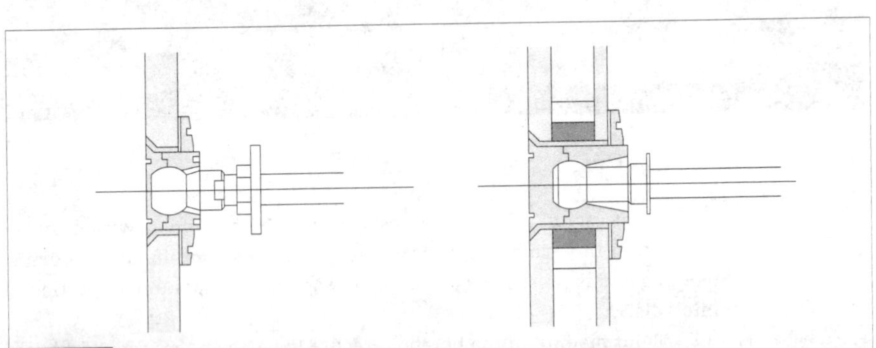
Fig.47:
Point mount with countersunk head for single glazing and insulation glass-pin-joint flush with the plate

Fig.48:
Facade with point mounts

in various forms: flush point mounts (flat head) and point mounts with attached pressure disks (clamp plates), for example, in lenticular (oval head) or cylindrical shapes. › Figs. 45–48

This system requires a precise static and structural plan, since the area of glass near the face of the hole is already damaged, and therefore weakened by drilling, and since this area also has to withstand the greatest tension under strain. The minimum spacing between boreholes or between boreholes and the exposed edge of the glass should not be less than 80 mm. For insulation glass, an additional edge seal is required around the drill hole to ensure the area between the glass panes remains watertight. Point mounts inside boreholes are elaborate and time-consuming to develop; it is common to use patented and approved production models for glass facades and roofs.

Undercut anchors

One special form of point mounting is the undercut anchor. This fastens the glass on one side only, and does not need to be bolted from the other side. The glass is fixed to the support by clamping a ductile cylinder in a conical borehole in the glass. › Fig. 49

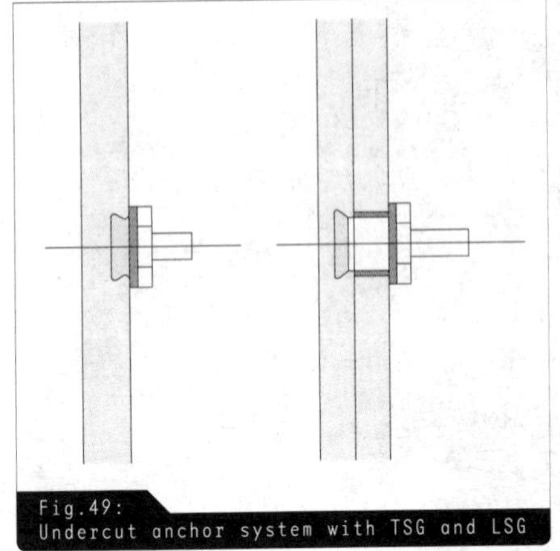
Fig.49:
Undercut anchor system with TSG and LSG

Fig.50:
Four-point joint

Point mounts with an integrated spherical joint are often used for glass facade structures, which helps avoid greater stress on the points caused by a bowing of the glass pane. Point mounts are fixed by stud bolts onto a substructure. Substructures and stud bolts should allow for subsequent settling of the glass panes and compensate for dimensional tolerance. A flush, steel profile with a slotted hole, for example, or a special four-point joint of stainless steel is appropriate for this system. › Fig. 50

### GLASS JOINTS AND GLASS CORNERS

Glass joints

If no cover molding is available, the glass edge of a point mount glazing is exposed. In a single glazing system with limited weather protection, such as a multistory parking garage or warehouse, the glass joints can remain open. Such facades are more cost-effective to build and also guarantee good ventilation. Yet it is necessary to seal the joints of insulation glass facades. The joints also have to be elastic, since glass expands from warming (dilatation) and facades can bow in the wind.

Sealing moldings

There are two basic different types of sealing profiles. The first involves pressing a sealing profile of EPDM/APTK or silicon into the joints. The second involves applying a spray sealant to the sealing profile, which seals the joints from the outside. › Figs. 51 and 52 In insulating or laminated glass, the rabbet should remain open to allow vapor pressure equalization and drainage, and thus guarantee the long-term durability of the edges of the glass. This method ensures that water from leakage or condensation does not remain too long in one place and can be quickly transferred outside via the drainage system. The exposed edge seal of the insulation glass must be UV resistant. A UV-resistant edge seal can be made by using

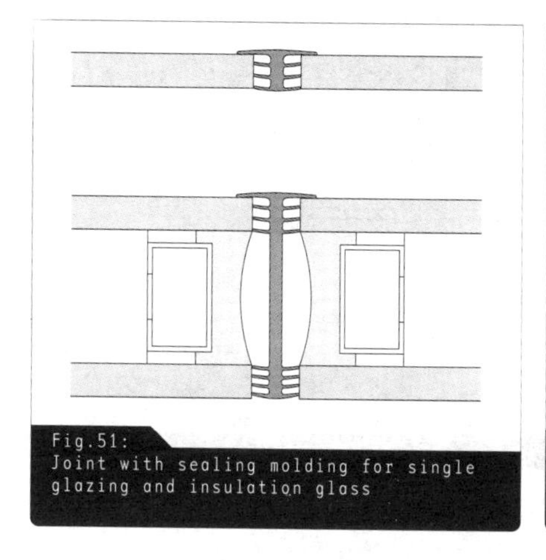

Fig.51:
Joint with sealing molding for single glazing and insulation glass

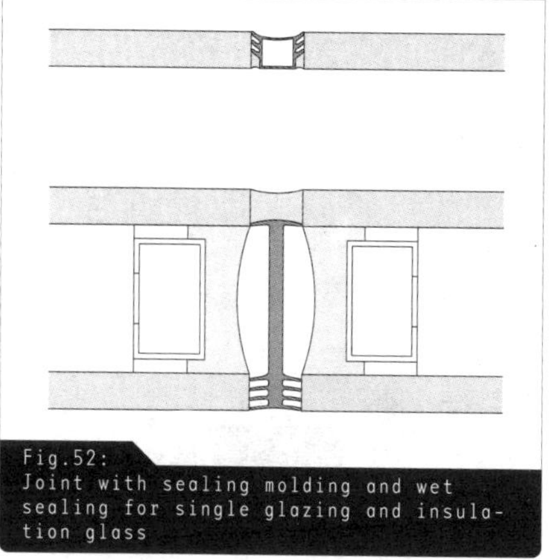

Fig.52:
Joint with sealing molding and wet sealing for single glazing and insulation glass

silicon (rather than polysulfide or polystyrene) in second-phase sealing, or by means an enamel strip, usually black, pressed into the inner side of the outer glass pane (Surface 2) and covering the edge seal.

Glass corners

The principle for glass corners is similar to that for glass joints. Vapor pressure equalization and the chemical compatibility of the sealant must be checked and guaranteed. The insulation qualities of glass corners are less favorable than the glazing itself, and condensation build-up should therefore be expected. There are diverse possible ways of constructing a glass corner with frameless glazing using insulation glass. The following solutions are the most common:

- Opaque corners: the corner is filled with an insulation molding › Fig. 53
- Stepped insulating glass with outer edges mitered › Fig. 54
- Stepped insulating glass, slotted together › Fig. 55

The forms for implementing stepped insulation, especially with a mitered edge, are more complicated and call for precise assembly and joints. The advantage, however, is that the view of the continuous glass surface is not interrupted by another material even in the corners.

### STRUCTURAL SEALANT GLAZING (SSG)

Structural sealant glazing constructions are systems where the glass pane is attached by adhesive. This special form of linear support is not automatically approved; therefore the chosen system requires individual authorization under building law. Gluing the glass panes onto a metal frame

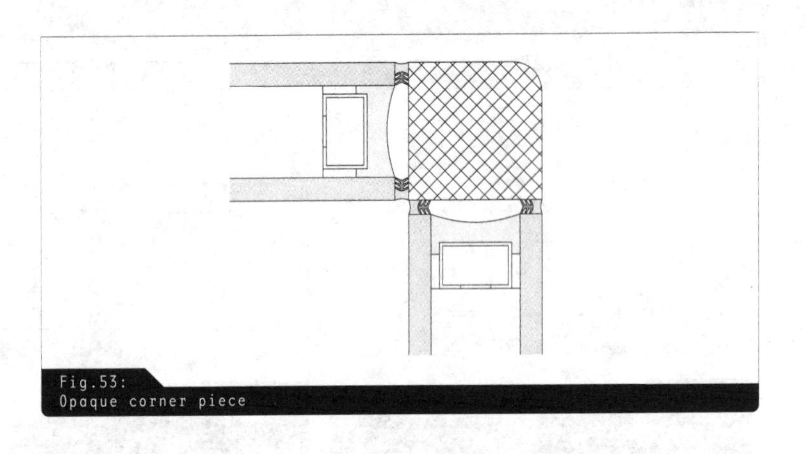
Fig.53:
Opaque corner piece

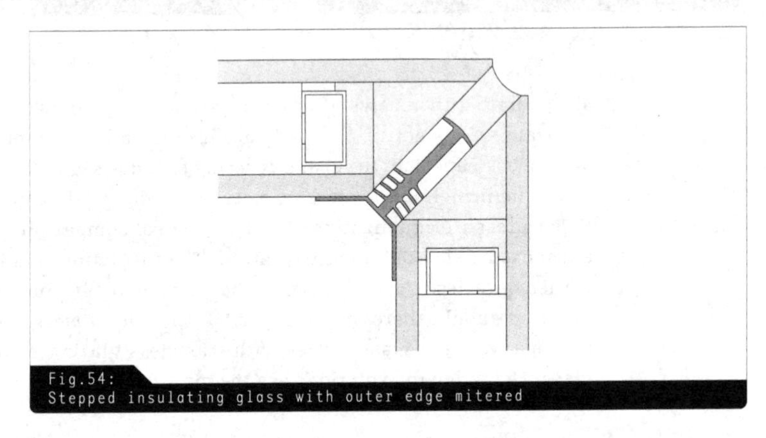
Fig.54:
Stepped insulating glass with outer edge mitered

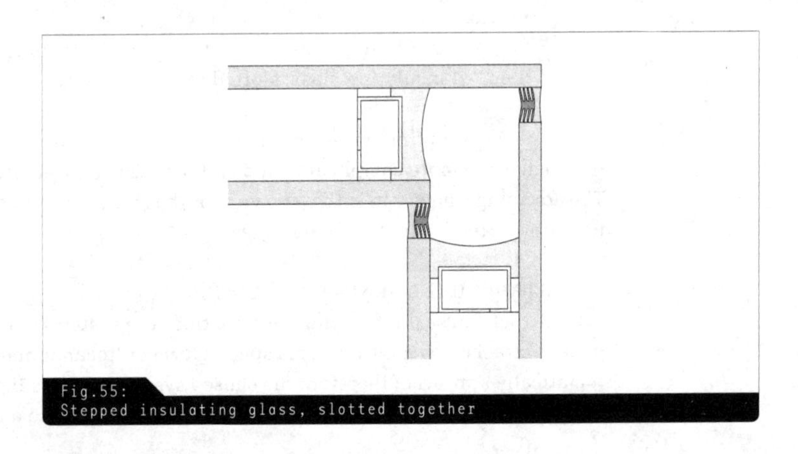
Fig.55:
Stepped insulating glass, slotted together

(adapter frame), which is also fixed to a support profile, provides a flush, frameless facade surface.

Bonding

Bonding is resistant to wind load; the weight of the glass is supported by the "classic" setting and location blocks system. All-around bonding cannot be carried out at the building site. It requires a licensed factory with a precise climate- and temperature-controlled environment that is also free of dust and debris. In most cases, bonding is done in the glass factory directly after the glass is produced. The glass panes must be absolutely clean, dry, and free of grease.

Suction safeguarding

In some countries such as Germany, glazing more than 8 m high requires an additional wind suction safeguard made of metal. It protects against falling glass if the bond fails. Great demands are made on the bond; it must withstand diverse loads, such as change in temperature, humidity, UV light, and possible corrosion due to microorganisms. The additional suction safeguard is required because it is very difficult to ascertain the long-term bearing capacity of the bond construction method. The suction safeguard can be mounted as an all-around frame, or point mounted. There are different systems for SSG facades for applying the bond, suction safeguards, joint sealant, and insulation glass, which is either step-mounted (stepped insulation glazing) or has a bluntly cut edge. › **Figs. 56–58**

The bond is compatible with silicon and polyurethane, among others. As is the case with point mounted glazing, SSG glazing also needs to be UV protected on the exposed edge bond. Bond constructions are increasingly being used for SSG glazing as well as all-glass structures. They also have the capacity to assume important structural functions. › **Chapter Applications**

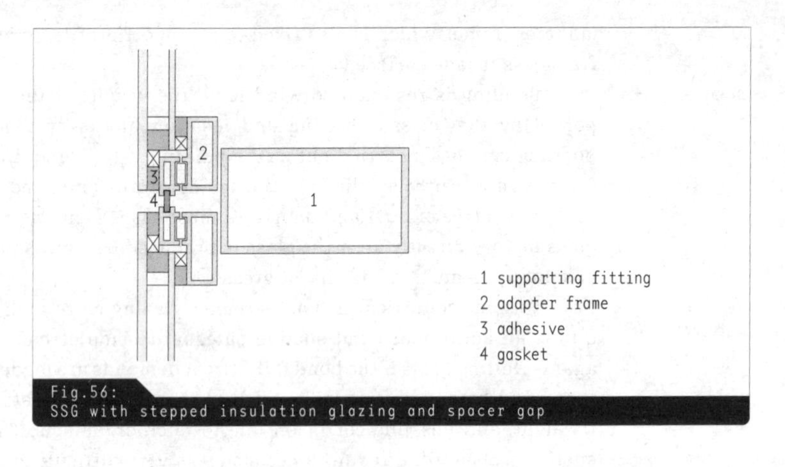


Fig.56:
SSG with stepped insulation glazing and spacer gap

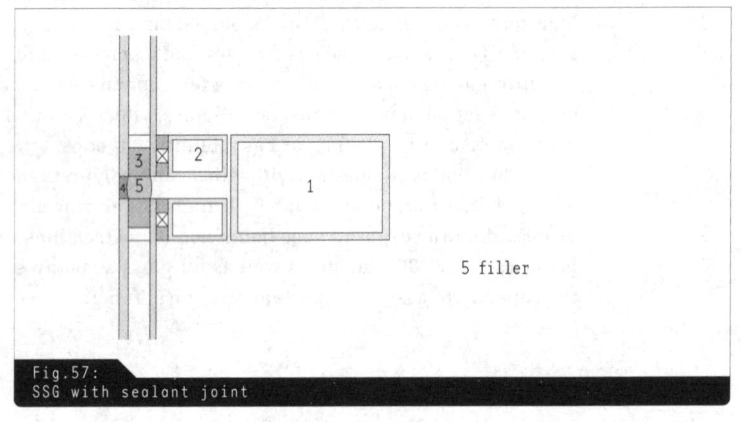


Fig.57:
SSG with sealant joint

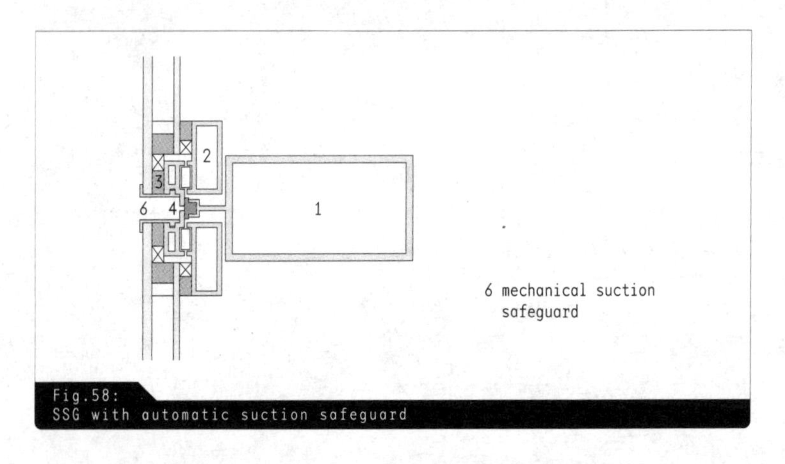


Fig.58:
SSG with automatic suction safeguard

## VERTICAL GLAZING

Whereas only certain types of glass are permitted for overhead glazing, › Chapter Construction and assembly theoretically any kind of glass can be used for vertical glazing. In practice, however, there are limitations based on the place it is installed, its use, and its construction.

Types of glass, risk of breakage

Toughened safety glass (TSG) or laminated safety glass (LSG) is often used in place of float glass to minimize the risk of accident. It is necessary to use safety glass, for example, in circulation areas in schools and kindergartens unless other measures, such as handrails or balustrades, have been taken to prevent people from colliding with the glass. Sometimes TSG is also used as an outer glazing for insulating glass to reduce the risk of glass breakage: this is particularly true of facades located above public circulation areas, or when the glass cannot be framed on all sides. For simple glazing that does not have linear mounts on all sides, the use of HSG, TSG, LSG, or if necessary wired glass, is recommended instead of float glass. In many countries this is even required. When glass is installed at a height above 4 m, toughened glass that has been heat soak tested is used instead of ordinary TSG to avoid spontaneous breakage from inclusions of nickel sulfide. › Chapter Glass as a building material

Point-supported glazing usually requires the use of tempered glass. When glass is installed at a height above 4 m, it is common to use laminated safety glass composed of two-pane TSG or two-pane HSG to prevent individual pieces from falling when the glass is damaged.

Load bearing, size of elements

Apart from design and functional criteria, the sizes of element are depend on the loads, the type of mounting, and the type of glass used. When calculating vertical glass elements, in addition to the weight of the elements themselves, wind load (wind suction and wind pressure), climate (for insulating glass › Chapter Glass as a building material), and especially traffic load (e.g. horizontal impact stresses in the case of store

\\Tip:
Glass with highly absorbent (solar control) coatings or body-tinted glass will be warmed by the sun more than neutral, colorless glass and should therefore be tempered. That applies above all to panes of glass that absorb more than 50% of incoming thermal radiation.

Fig.59:
Installing a large pane of insulating glass at a construction site

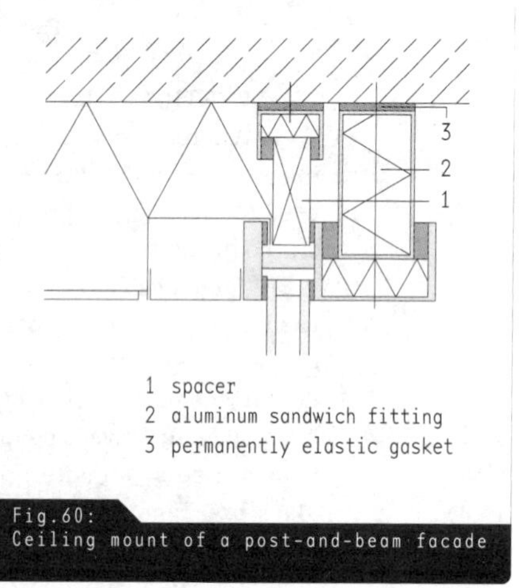

Fig.60:
Ceiling mount of a post-and-beam facade

windows) must be taken into account. Theoretically, using linear mounting on all sides, single-pane glazing and even insulating glazing as large as 3.21 × 6.00 m (ribbon size) could be installed. For linear mounting on fewer than four sides or for point mounting, the largest possible formats are considerably smaller than that, since they have to absorb greater bending and higher tension in the glass. The weight of the glass limits size even further: a pane of insulating glass von 3 × 6 m in size can easily weigh as much as 2 t. The installation of large and heavy plates of glass is complicated and requires heavy-duty vacuum lifters to hoist the glass. › Fig. 59

Facade types

In essence, two types of glass facades are distinguished: post-and-beam facades and modular facades.

The term post-and-beam refers to the bearing structure of the facade, which consists of vertical elements (posts) and horizontal ones (beams). The individual posts and beams are assembled on site by welding or screwing them together. Then the glass is fastened to this substructure from outside. The post-and-beam facade makes it possible to build large spans but has the disadvantage that the onsite assembly is more complicated and takes more time than modular facades require.

Modular facades consist of elements prefabricated in the factory into which all the structural elements, such as frame, glazing, and casements, are already integrated. › Chapter Applications, Openings The dimensions of the prefabricated elements have to be suitable for transportation, which limits the size of the elements available.

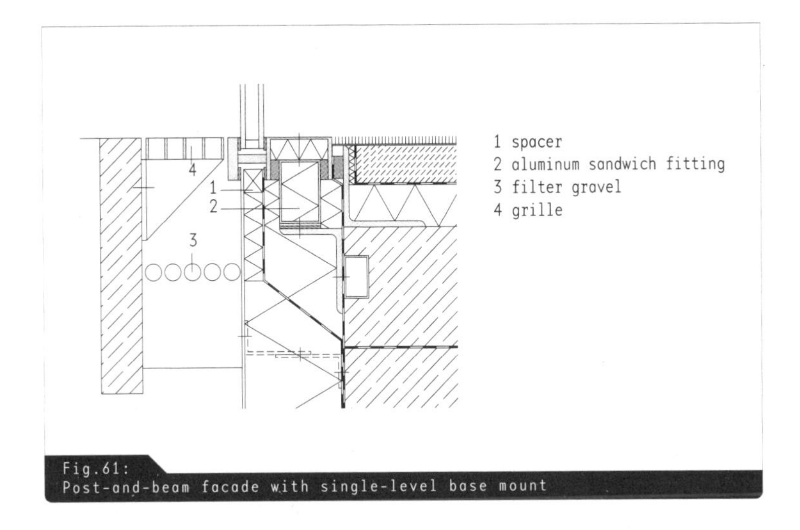


Fig. 61:
Post-and-beam facade with single-level base mount

Mounting details

The mounts of a glass for attaching a glass facade to adjoining building parts, such as the roof or floor structure, have to be constructed in a way that they do not cause additional load on the glass panes in the glazing unit.

Roof mount, ceiling mount

For that reason, the facade in the glazing unit is attached to the ceiling using a plastic spacer or an aluminum sandwich fitting. Traffic load or movements of the roof caused by settling are stabilized by such flexible mounts. The seal and the thermal separation of the facade are preserved. › Fig. 60

Base mount

A spacer is also used for the base mount of a post-and-beam facade; it is attached together with waterproofing to the lower beam of the facade and fastened behind the pressure molding. It is important that the point where the seal and facade are attached be at least 15 cm above the layer that conducts the water. With a single-level mount on the exterior, a gutter is vital to ensure drainage. › Fig. 61

## OPENINGS

Openings in a building's shell have essential functions to fulfill. Moreover, their design is especially important for the overall look. In the case of glass facades, openings are also important because they help regulate the climate in the building. They permit natural ventilation of the interior and thus help keep solar radiation from continually warming of the air. In addition to classic casement windows, additional opening elements are employed in glass facades and glass roofs to ensure continuous

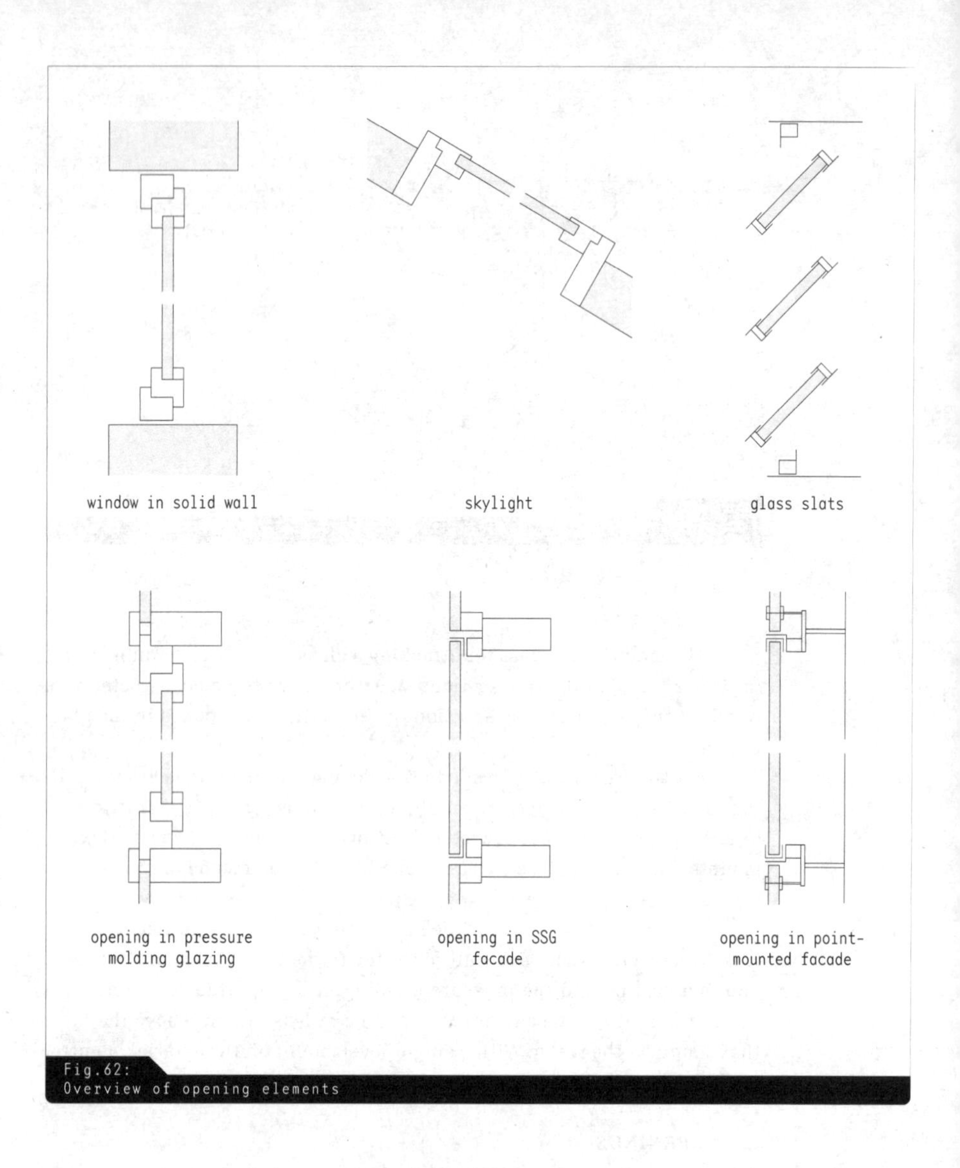


Fig.62:
Overview of opening elements

ventilation; they can be automatically controlled according to room temperature. Various types of openings in glass facades are listed and described below. › Fig. 62

The window element in a wall is the simplest and the original form of glass facade. It can be composed of a single field or be divided into several, which can in turn have either fixed glazing or panes that open (sashes).

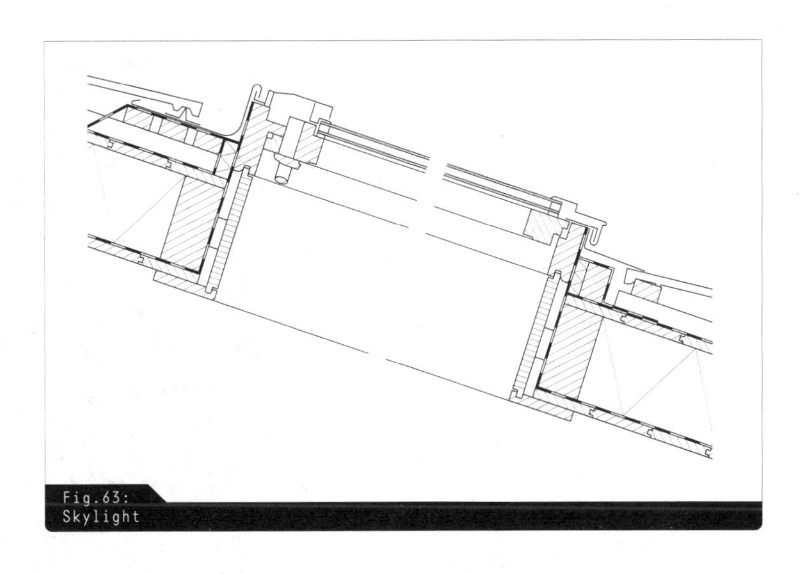

Fig. 63:
Skylight

The skylight provides light and ventilation for rooms under ceilings. The glass has to meet the requirements for overhead glazing. The skylight should always be installed at an angle to ensure unrestricted water drainage. › Fig. 63

Glass slats or slat windows make it possible to set precisely the required ventilation cross section and hence regulate the ventilation of the room. Slats are available in various types, with and without frames, for single-pane and insulating glass. › Figs. 64 and 65

The openings discussed above can also be integrated into pressure molding glazing. The glass of the window frame of the opening element is replaced by a supporting frame and pressure molding, which are then clearly visible from the outside.

\\Tip:
More on the theme openings may be found in Roland Krippner and Florian Musso, *Basics Facade Apertures*. Basel: Birkhäuser Verlag, 2008.

\\Tip:
Slats of single-pane glass are used for adjustable, exterior sunshades. The surface of the glass is printed or coated for that purpose. With the addition of integrated solar cells, electricity can be produced to operate the glass slats.

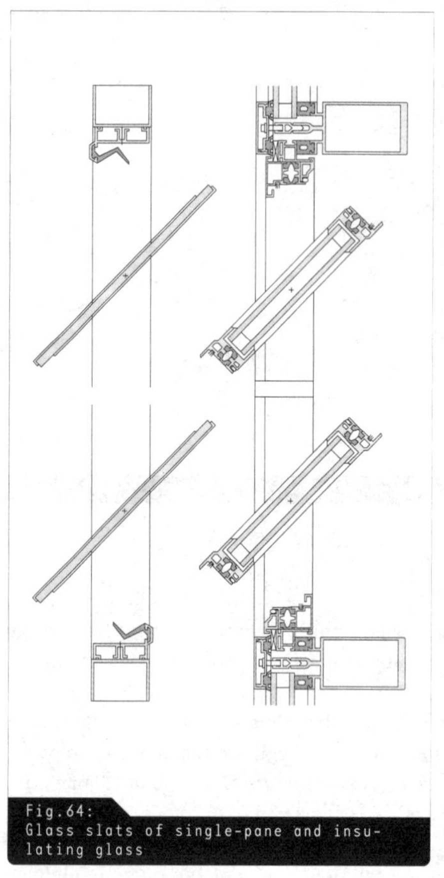

Fig. 64:
Glass slats of single-pane and insulating glass

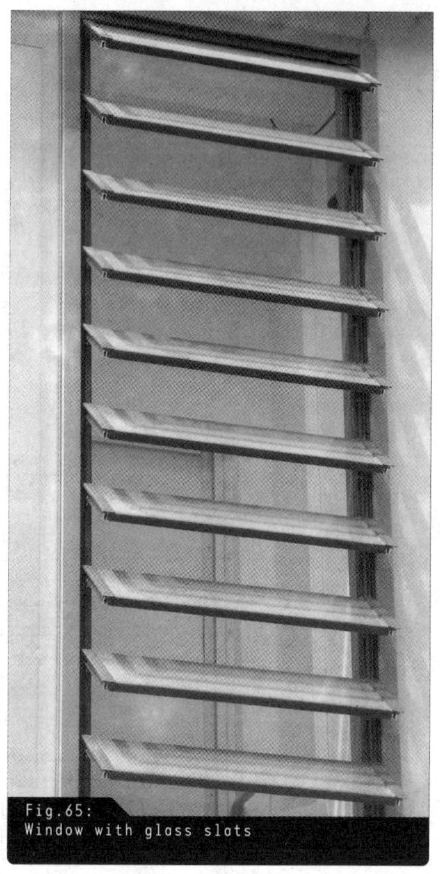

Fig. 65:
Window with glass slats

For openings in SSG facades, top-hung sashes that open outward are used. This solution provides for openings with frameless views outward, which are very attractive in SSG facades. › Fig. 66

The integration of opening elements in point-mounted glazing is a special challenge, since the filigree appearance of the facade should not be disturbed in the area of the windows either. For facades with single-pane glazing, for example, top-hung sashes or slats of point-mounted glass can be employed.

Facades of insulating glass require considerably more effort. Because of the lack of frame, the possibilities for attaching seals and fittings that will provide sufficient thermal insulation are very limited. It is possible, for example, to install top-hung sashes as on SSG facades.

Fig. 66: Openings in SSG facade

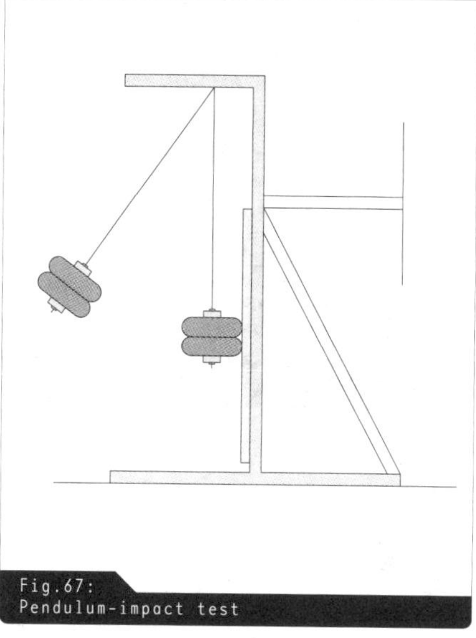
Fig. 67: Pendulum-impact test

**FALL-PREVENTION GLAZING**

Glass structures used in place of a banister or balustrade required under the building code to prevent people from falling are called fall-prevention glazing. The spectrum of applications ranges from balustrade infill of glass by way of all-glass balustrades to glazing the full height of the room with linear or point mounts. In many countries, the security of fall-prevention glazing has to be demonstrated by a dynamic stress test such as the pendulum-impact test. › Fig. 67

Because such testing is quite complicated, the structure should be planned so that experience from tried-and-tested constructions is applied and further stress testing can be avoided. Glazing that prevents falls can be divided into three categories:

Fixed full-height glazing

The first is fixed full-height glazing—that is, glazing without opening sashes, handrails, or protruding rails to support horizontal loads. Linear mounts on all sides are the most effective solution from a structural perspective. If this is not possible, the free edges of the glazing have to be protected from impact in other ways, such as adjoining panes of glass or neighboring architectural elements such as walls or ceilings. Panes of glass mounted on two sides will bend considerably more in case of impact because two edges are free. Care should therefore be taken to inset the glass sufficiently, so that the pane cannot slide out of the holder. In the

Fig.68: All-glass balustrade

Fig.69: Banister with point-supported glass infill

case of point mounting, the clamp plates must have a diameter of at least 50 mm.

Fixed, bearing balustrade

The second category includes bearing glass balustrades that have linear mounts on their bottom edge with a clamp construction. The upper edge should be protected against impact, for example, with a fitting applied with adhesive or a continuous mounted handrail. The latter should be of appropriate size to ensure that if one pane fails the horizontal load will be transferred to the next pane. The appropriate type of glass for this category is LSG made of HSG or TSG. › Fig. 68

Balustrade infill

The third and last category is balustrade glass in facades and glass banisters in which glass is used as infill. The horizontal loads are supported by a bearing handrail or a crossbar (facade). The glazing has either point mounts or linear mounts on at least two opposite sides. The panes should be either LSG or TSG. › Fig. 69

\\ Note:
In most cases, fall-prevention glazing calls for the use of laminated safety glass, and in the case of point mounting, LSG made from HSG or TSG. If the pane is broken, the laminate of glass and highly tear-resistant film still offers sufficient protection for the person colliding with it.

\\ Tip:
One special case is fixed full-height glazing with a bearing rail whose height is specified in the building code, for example, a round stainless steel fitting on the inside, which will largely absorb any stress from impact. Here, the thickness of the glass can be reduced without altering its outward appearance.

## OVERHEAD GLAZING

Overhead glazing, like vertical glazing, can have either linear or point mounts.

Bearing loads, dimensions

Structural loads on overhead glazing is higher than on vertical glazing, since the pane's own weight runs perpendicular to its plane, and because wind, climate, and snow loads have to be taken into account as well. Because of these increased stresses, the format of the elements cannot be as large in overhead areas as on facades. For larger glass roofs, additional loads for people and material have to be assumed as well, since it has to be possible to walk on the glazing for maintenance and cleaning. › **Chapter Applications, Glazing for restricted and unrestricted foot traffic**

In addition, a glass roof should be resistant to impact, which can result from hailstones or falling branches, for example. Tempered glass should therefore be used as the upper layer on a glass roof. For point-mounted overhead glazing, the required thickness depends not only on loads but also on residual bearing capacity. The same is true of the diameter of the clamp mounts. › **Tab. 7**

Tab.7:
Point-mounted overhead glazing with certified residual bearing capacity with rectangular grid of supports

| **Diameter of clamp mounts (mm)** | **Minimal thickness of glass (mm) LSG made of HSG** | **Spacing of supports (cm) in direction 1** | **Spacing of supports (cm) in direction 2** |
|---|---|---|---|
| 70 | 2 × 6 | 90 | 75 |
| 60 | 2 × 8 | 95 | 75 |
| 70 | 2 × 8 | 110 | 75 |
| 60 | 2 × 10 | 100 | 90 |
| 70 | 2 × 10 | 140 | 100 |

\\ Note:
In the case of glass enclosures of elevator shafts or escalators, additional measures are usually necessary to ensure the safe operation of the conveyor system. For example, the glazing should be installed in such a way that it is not possible to reach over or between the plates of glass (in the case of frameless glazing).

\\ Note:
For point mounted overhead glazing, the minimum requirement is LSG consisting of 2×6 mm thick HSG and PVB film at least 1.52 mm thick. The diameter of the clamp plates should be at least 60 mm. The free edge of the glass–that is, the distance between the edge of the glass and the point mount–has to be at least 80 mm and may be at most 300 mm.

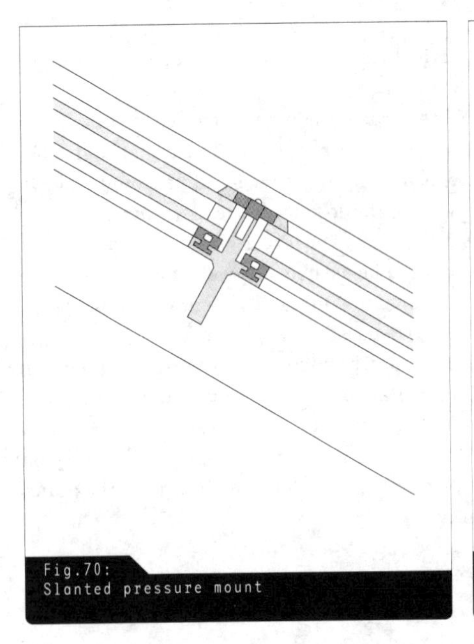
Fig.70:
Slanted pressure mount

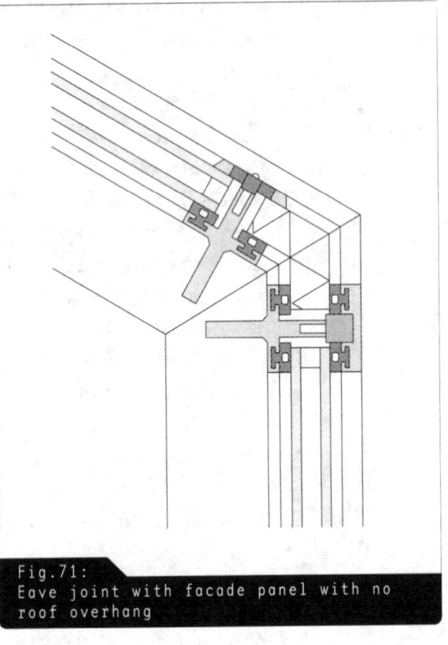
Fig.71:
Eave joint with facade panel with no roof overhang

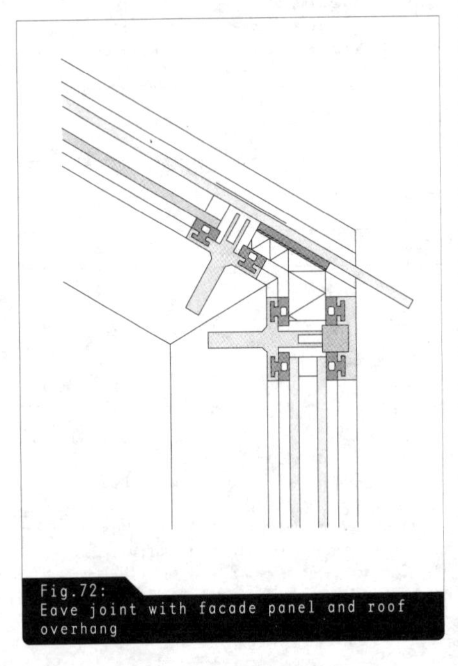
Fig.72:
Eave joint with facade panel and roof overhang

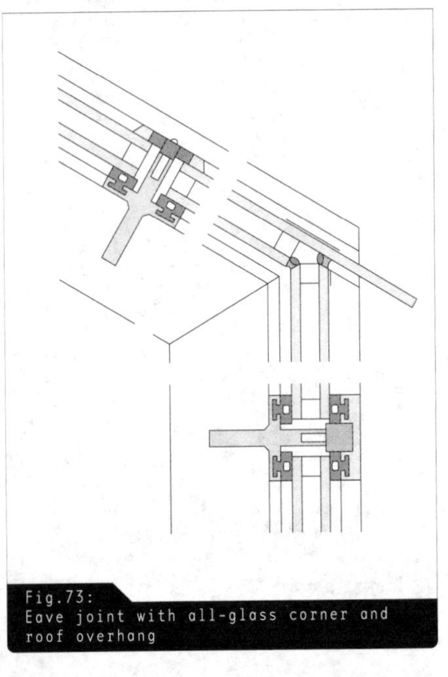
Fig.73:
Eave joint with all-glass corner and roof overhang

Roof drainage

The glass surface should always have sufficient slope leading to the level where water is directed, that is, to a drainage gutter or an adjoining flat roof. Pressure moldings (pressure molding glazing) should installed at an angle so the water can run off. › Fig. 70

The eave joint with attachment to a vertical glazing can be constructed with or without an overhang. If there is no overhang, the rainwater runs directly down the facade, dirtying it. Larger glass roofs should therefore be drained via gutters. › Figs. 71–73

## GLAZING FOR RESTRICTED AND UNRESTRICTED FOOT TRAFFIC

Glazing for restricted foot traffic

Overhead glazing that has to be stepped on for maintenance and cleaning is referred to as step-on glazing. It should be noted that, despite the name, only a limited number of maintenance people can step on the glass at the same time. In addition to the structural strength of the glass, it is necessary to plan measures to avoid slipping. If the slope of the roof is greater than 20°, safety hooks should be installed. The glazing itself should be LSG with certified impact resistance and residual bearing capacity. With insulating glass, the upper pane can be TSG rather than LSG.

Glazing for unrestricted foot traffic

Glazing for unrestricted foot traffic can, unlike glazing for restricted foot traffic, also be accessed by the public, and hence it is subject to considerably higher traffic load. Usually a maximum of 5.0 $KN/m^2$ is assumed. Walk-on glazing typically employs LSG approx. 30 mm thick or more. The glazing is often composed of three individual panes, with the upper pane of TSG or HSG to provide an impact-resistant surface and to protect the two lower panes from damage. The bearing function is fulfilled by the two lower panes, which is why the glazing will remain strong enough for standing on even if the covering pane is damaged. › Fig. 74

Skid-proofing

The surface should be made skid-proof, either by means of special ceramic screen printing that offers a rough surface on all or part of the glass. A rough, skid-proof surface can also be produced by etching. Glazing that will be walked on usually has linear mounts on two or four sides. In the former case, it must also be bolted to the supporting structure. With an insert of at least 30 mm, panes of glass rest on pressure-resistant elastomer supports. Glass-to-glass or glass-to-metal contact on the sides is prevented by inserting spacers. Glazing suitable for foot traffic is used, for example, for interior stairs of floor structures. Construction solutions for insulated glass suitable for foot traffic are complicated, since high traffic loads cannot be shifted to the edge assembly, as it can with glass suitable for light traffic, since the edge assembly would be damaged by the constant strain. Glass suitable for light traffic, which is used for example to illuminate exhibition areas in museums, is therefore often

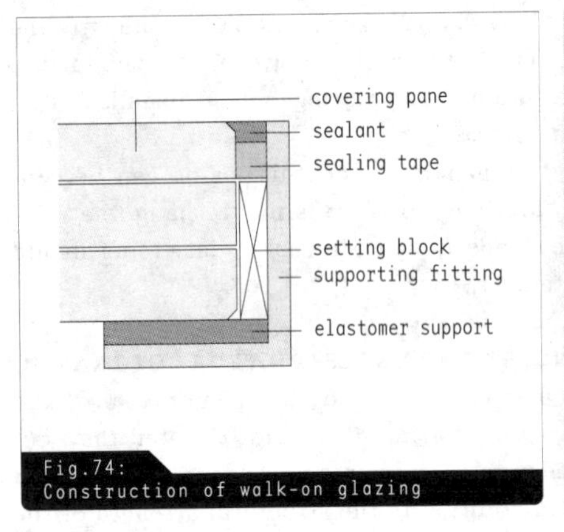

Fig.74:
Construction of walk-on glazing

Fig.75:
Glazing for unrestricted foot traffic

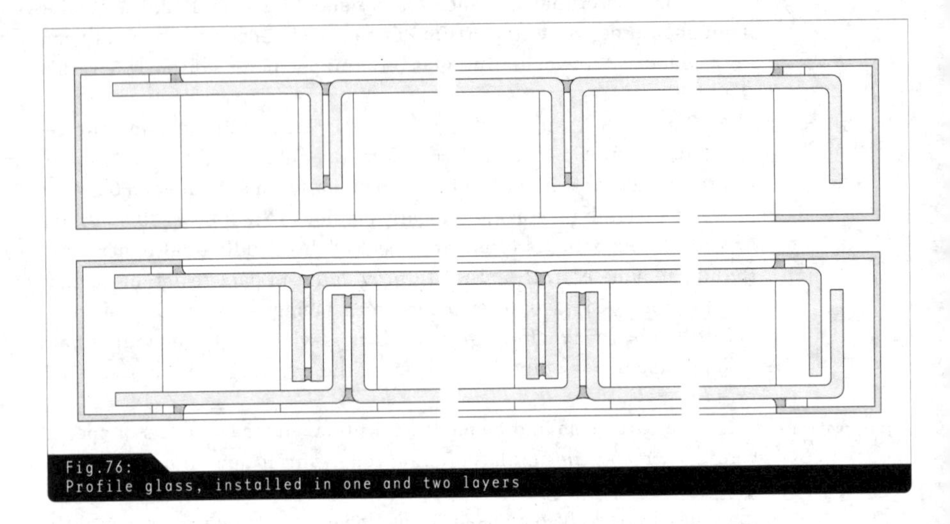
Fig.76:
Profile glass, installed in one and two layers

constructed as a two-layer roof. The insulating glass required for thermal separation is thus on the second layer beneath the glass suitable for light traffic. › Fig. 75

### PROFILE GLASS

The advantage of profile glass is that there is no need for a substructure on the facade to bear the load. Rigid glass profiles make it possible to install glass across large surfaces without using muntins. Profile glass

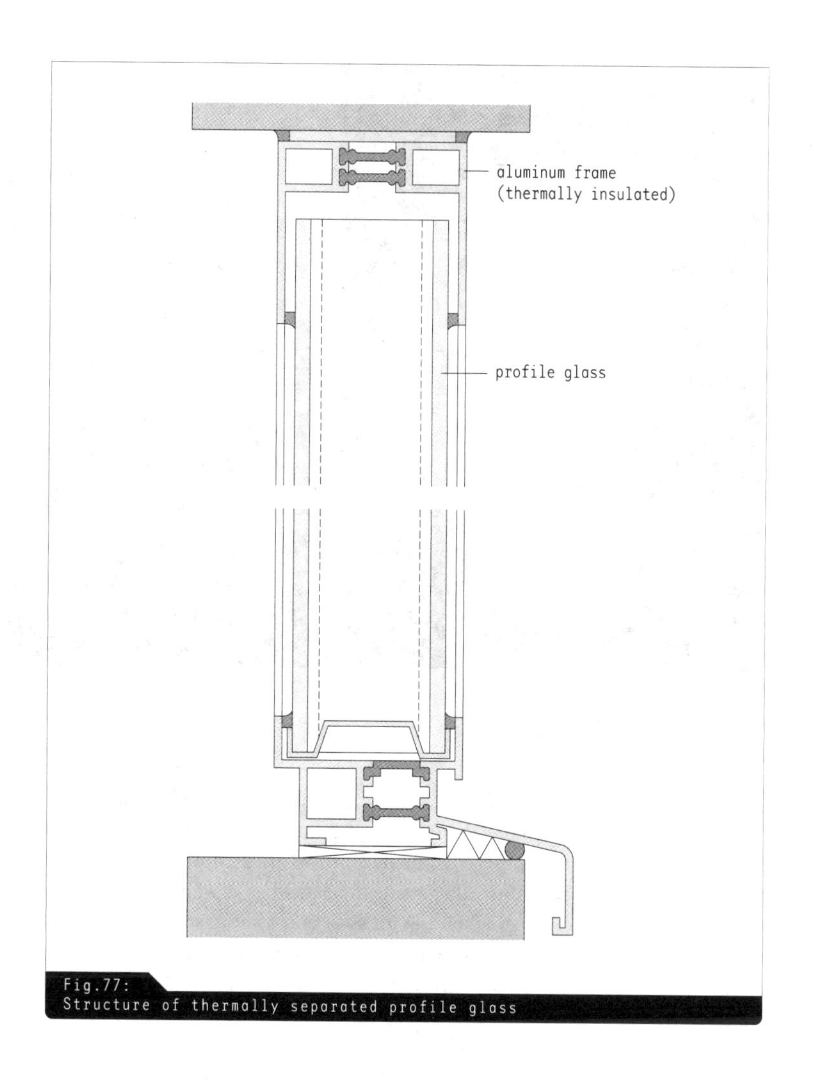


Fig.77:
Structure of thermally separated profile glass

can be installed vertically or horizontally. For purposes of attachment and load transfer, the profiles are inserted into the narrow sides of an approx. 50 mm deep aluminum frame. They can be installed in one or two layers. › Fig. 76

Thermal separation

A thermally separated glass facade is only possible with two-pane glass. The glass profiles are then installed facing one another. Then setting blocks and sealant are applied to the joints. With a two-layer facade, it also makes sense to have a thermally separated frame. › Fig. 77

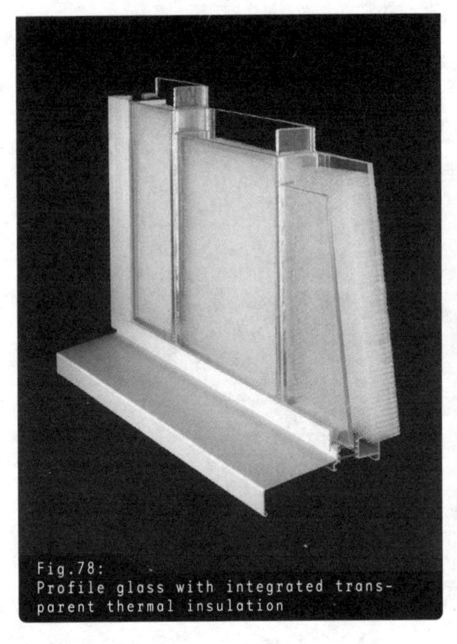
Fig.78:
Profile glass with integrated transparent thermal insulation

Fig.79:
Example of facade with profile glass

Because the space between the glass profiles cannot be dehumidified as it can with insulating glass, an opening to the outside has to be planned to avoid condensation by permitting the damp air to escape. Profile glass can be tempered, which improves its structural characteristics and safety. Tempered profile glass can be as long as approx. 7 m. In addition, wire can be rolled in when the glass is manufactured to provide protection against splintering. Such profiles cannot, however, be tempered.

Surface design, coating

Various kinds of surface designs can be manufactured on profile glass, from smooth to patterned. As a result of the manufacturing process, however, the surface of smooth profile glass is not as even as that of float glass, which clearly detracts from the quality of its transparency. › Fig. 80 Profile glass is now available in coated versions as well, for example, with coatings for thermal insulation or solar control. On a two-layer glazing with profile glass, the coating for thermal insulation is placed on the inner layer, while the coating for solar control is on the outer layer. It is also possible to integrate transparent thermal insulation into profile glass. › Figs. 78 and 79

This is done by placing a capillary insert in the air space between the panes to scatter natural light. This effect is used above all for buildings that will benefit from even, glarefree distribution of light, such as sports arenas, workshops, museums, and studios.

Fig. 80: Transparent profile glass

Fig. 81: Glass supporting structure with linearly mounted glass facade

Fig. 82: Glass supporting structure with point-mounted glass facade

Fig. 83: Supporting grid of glass

## GLASS SUPPORTING STRUCTURES

The dematerialized structures of our age reflect the technical advances of recent decades concerning glass as a building material. Glass in buildings is no longer limited to serving as a shell but has begun to take over supporting functions. Glazing for unrestricted foot traffic has already been introduced, but its substructure is not usually made of glass. Since the 1990s, however, glass buildings have been building employing glass as the primary building material for the supporting structure. That means that the supports or posts of a glass facade, which provide reinforcement and absorb wind loads, are also made of glass.

Beams of flat glass

Rather than steel or wooden profiles, slender blades, for example, of laminated safety glass form the supporting structure. As with facades, the supporting members of a roof can also be made of glass. This ensures maximum transparence and lighting of a covered courtyard or interior. › Figs. 81–83

Fig.84:
Glass footbridge connecting two buildings

Fig.85:
Detail of the footbridge

Increasingly innovative constructions are constantly expanding the number of all-glass buildings. For example, there are glass connecting bridges, all-glass stairways, experimental supports of glass tubes, and arch and shell supporting systems. › Figs. 84–87

Bearing glass tubes

Tubes of borosilicate glass have a cross section with good structural properties and are particularly well suited to absorbing high compressive forces. They can replace supports of concrete, steel, or wood in buildings. Particular attention must be paid to the steel parts on the ends of the tubes, which have to distribute the forces evenly over the cross section of the tube. The load is transferred to adjoining parts of the building by means of a ball joint, which prevents the shear forces and bending moments from being transferred to the cross section of the glass.

Shell and arch supporting structures

Especially for supporting structures that are primarily subject to compressive forces and not bending moments, such as supporting structures in the form of arches, domes, or shells, glass, with a compressive strength much higher than its tensile strength, can demonstrate its capabilities.

Calculating glass supporting structures

The prerequisites for the development described were, first, new possibilities of processing sheet glass into TSG, HSG, or multilayer LSG, which considerably improved the bearing and residual bearing capacity of such

Fig.86:
Glued glass stair construction

Fig.87:
Detail of glued stair construction

constructions. Second, this progress in the most recent methods was due to the predetermination of the bearing behavior of glass through experiment and calculation. In contrast to ductile materials like steel, which can reduce high tensions by distorting plastically, points of high tension will cause glass to break spontaneously. The load limit of a specific pane of glass is difficult to determine, because it depends on how much it has already been damaged (scratches, small breaks on the edges). The tensions permitted are therefore often much lower than the actual load limit of the pane. This ensures sufficient protection against spontaneous breakage, but it also means that all-glass constructions cannot be quite as delicate as would theoretically be possible.

Connectors

The particular difficulty of all-glass constructions is usually not redirecting forces in the sheet of glass itself but rather transferring them from one component to the next. In the region of such transfer points, the joint is either made by means of classical mechanical connectors (point or clamp mounts) or through adhesive connections. Adhesive connections have the advantage that they permit a uniform (and hence suited to the material) distribution of load to the glass. The transfer of load to the adhesive surfaces can, however, be diminished by outside influences such as dampness, temperature, or aging. In practice, therefore, silicon adhesives are usually employed, although research is currently being conducted on other suitable adhesives with greater strength, such as hot melt adhesive foils.

Glass beams are made of laminated glass, usually composed of three or more panes of TSG or HSG. The outer panes provide protection, while the inner ones provide the actual support. At present, there are no universal regulations for such constructions, so that the building codes differ greatly from country to country.

The technological limits of all-glass structures are determined not only by the strength of the materials but also by the possibilities for manufacturing them. Most processing plants can produce laminated safety glass up to 7 m long, and as thick as 80 mm. Greater lengths require an autoclave capable of handling special sizes, something few companies have.

## IN CONCLUSION

The visions of Bruno Taut (Haus des Himmels [House of the Sky], 1920) or Ludwig Mies van der Rohe (glass high-rise project, Berlin 1921) testify to the fascination that glass has for architects. Nearly ninety years later, the building material of modernism has not ceased to be a modern building material.

Because of climate change, the ecological responsibility of architects is greater than ever today. At least since energy consumption of buildings has been restricted, the new types of glass with effective thermal and solar control are key components for creating thermal shells of glass. With the help of computer programs for the realistic simulation of how buildings respond to the climate and knowledge gained from experience with existing buildings, glass facades can be part of an appropriate energy concept. In recent years, new aspects have come to the form in the area of design as well. Dematerialization and transparency are no longer the only objectives that dominate glass architecture. "Materialization" in the form of colored or translucent glass has become a popular means of design, and a variety of products are available.

In all likelihood, such technical innovations as switchable layers, vacuum insulating glass, and self-cleaning surfaces will play more important roles in the future. It is already clear today that the evolution of glass technology is by no means over.

The possibilities of this building material, which continues to open new perspectives on its use and design, makes glass a material with the potential to remain an interesting and exciting field in the future as well.

# APPENDIX

## STANDARDS, GUIDELINES, REGULATIONS

| **Glass as material** | |
|---|---|
| EN 572 | Glass in building–Basic soda lime silicate glass products |
| EN 1051 | Glass in building–Glass blocks and glass pavers |
| EN 1096 | Glass in building–Coated glass |
| EN 1279, parts 1-6 | Glass in building–Insulating glass units |
| EN 1863 | Glass in building–Heat strengthened soda lime silicate glass |
| EN 12150 | Glass in building–Thermally toughened soda lime silicate safety glass |
| EN 14179 | Glass in building–Heat soaked thermally toughened soda lime silicate safety glass |
| EN ISO 12543 | Glass in building–Laminated glass and laminated safety glass |
| ASTM C1048-04 | Standard Specification for Heat-Treated Flat Glass–Kind HS, Kind FT Coated and Uncoated Glass |
| ASTM C1172-03 | Standard Specification for Laminated Architectural Flat Glass |
| ASTM C1376-03 | Standard Specification for Pyrolytic and Vacuum Deposition Coating on Flat Glass |
| ASTM C1464-06 | Standard Specification for Bent Glass |

| **Thermal protection** | |
|---|---|
| EN 673 | Glass in building–Determination of thermal transmittance (U value) |

| **Sun protection** | |
|---|---|
| EN 410 | Glass in building–Determination of luminous and solar characteristics of glazing |
| ISO 9050: 2003 | Glass in Building–Determination of light transmittance, Solar direct transmittance, total solar energy transmittance, ultraviolet transmittance and related glazing factors |

| **Security** | |
|---|---|
| EN 356 | Glass in building-Security glazing-Testing and classification of resistance against manual attack |
| EN 1063 | Glass in building-Security glazing-Testing and classification of resistance against bullet attack |

| **Fire protection** | |
|---|---|
| EN 357 | Glass in building-Fire resistant glazed elements with transparent or translucent glass products-Classification of fire resistance |
| EN 13501-1 | Fire classification of construction products and building elements-Part 1: Classification using data from reaction to fire tests |

| **Statics, stability** | |
|---|---|
| EN 13022 | Glass in building-Structural sealant glazing |
| EN 13474 | Glass in building-Design of glass panes |
| ASTM E2358-04 | Standard Specification for the Performance of Glass in Permanent Glass Railing Systems, Guards and Balustrades |
| ANSI Z97.1-2004 | Approved American National Standard-Safety Glazing Materials used in Buildings-Safety Performance Specifications and Methods of Tests |
| EOTA | Guideline for European Technical Approval for Structural Sealant Glazing Systems (SSGS) |

**LITERATURE**


Achilles, Andreas. "Coloured Glass: Manufacture, Processing, Planning." *Detail: Review of Architecture and Construction Detail* 2 (2007): 184–87.

Button, David, and Brian Pye, eds. *Glass in Building*. Oxford 1993.

Compagno, Andrea. *Intelligente Glasfassaden/Intelligent Glass Façades: Material Anwendung Gestaltung/Material, Practice, Design*. 5th ed. Basel: Birkhäuser Verlag, 2002.

Krippner, Roland, and Florian Musso. *Basics Facade Apertures*. Basel: Birkhäuser, 2008.

Kruft, Hanno-Walter. *A History of Architectural Theory: From Vitruvius to the Present*. London: Zwemmer; New York: Princeton Architectural Press, 1994.

Rice, Peter, and Hugh Dutton. *Structural Glass*, London; New York: E & FN Spon, 1995.

Schittich, Christian, ed. *Building Skins*. Munich: Edition Detail; Basel: Birkhäuser, 2006.

Sobek, Werner. "Glass Structures." *The Structural Engineer* 83/7 (April 2005).

Staib, Gerald, Dieter Balkow, Matthias Schuler, and Werner Sobek. *Glass Construction Manual*. Munich: Edition Detail, 2006.

Weller, Bernhard, and Thomas Schadow. "Structural Use of Glass." *Detail: Review of Architecture and Construction* 2 (2007): 188–90.

Wurm, Jan. *Glass Structures: Design and Construction of Self-Supporting Skins*. Basel: Birkhäuser Verlag, 2007.

## PICTURE CREDITS

Figs. 2, 4, 8, 17, 18, 19, 20, 21, 22, 29, 31: photographer: Martin Lutz (Akademie der Bildenden Künste), Stuttgart; copyright: Andreas Achilles, Jürgen Braun, Peter Seger, Thomas Stark, Tina Volz, Stuttgart

Fig. 3, figs. 24, 25 (Cineplexx Salzburg); figs. 26, 27, 78; fig. 79 (IHK Würzburg); fig. 80 (Uni-Klinik Hamburg): Glasfabrik Lamberts GmbH & Co. KG, Wunsiedel

Fig. 23: Saint-Gobain Glass Deutschland GmbH, Aachen

Fig. 28 (Landeszentralbank Meiningen): Schott AG, Mainz

Fig. 30 (Technologie und Innovationszentrum Grieskirchen); fig. 66 (Douglasgebäude Linz); fig. 82 (Palais Coburg, Vienna): Eckelt Glas GmbH, Steyr

Fig. 36: Schüco International KG, Bielefeld

Figs. 37, 38, 40, 41, 42: Institut für internationale Architektur-Dokumentation GmbH & Co. KG, Redaktion Detail, Munich

Fig. 44 (Kunsthaus Bregenz): Hélène Binet, London

Fig. 59: Siegfried Irion, Stuttgart

Fig. 83 (Technische Universität Dresden): Institut für Baukonstruktion, Technische Universität Dresden, Dresden

Figs. 84, 85 (Glass bridge, Schwäbisch Hall): Glas Trösch Beratungs-GmbH, Ulm-Donautal

Fig. 86 (Glasstec Düsseldorf): René Tillmann, Düsseldorf

Fig. 87: Andreas Fuchs, Universität Stuttgart IBK Forschung und Entwicklung, Stuttgart

Figs. 1, 5, 6, 7, 9, 10, 11, 12, 13, 14, 15, 16, 32, 33, 34, 35; fig. 39 (LBBW highrise, Stuttgart); fig. 43 (Kronen Carré, Stuttgart); figs. 45, 46, 47; fig. 48 (office building on Königstraße Stuttgart); figs. 49, 50, 51, 52, 53, 54, 55, 56, 57, 58, 60, 61, 62, 63, 64, 65, 67, 68, 69, 70, 71, 72, 73, 74, 75, 76, 77; fig. 81 (Kunstgalerie Stuttgart): Andreas Achilles, Diane Navratil, Stuttgart

## THE AUTHORS

Andreas Achilles, engineer, was a lecturer at the Institut für Baukonstruktion und Entwerfen at the Universität Stuttgart, and is now a freelance architect and author based in Stuttgart.

Diane Navratil, engineer (architecture and urban planning), works in the urban planning department in Karlsruhe.

P10

# 导言

玻璃不同于多数材料，在功能性之上，它更具有符号意义，并令人着迷。哥特时期的玻璃窗就已经开始刻意渲染光线使人产生超越尘俗之感。在现代主义建筑的视野中，不论在何种潮流中产生了何种具体的玻璃应用手法，玻璃这通透的材料本身总是至关紧要的。玻璃之重要，不仅仅在于它通透得可以产生近乎无形的外观，并带来流动性的开放空间；其重要还在于优雅、棱角分明、绚丽闪烁的特质。时间终将证明，将玻璃从填充小小窗户的嵌板角色解放出来，并使其成为独立要素的变革是多么高瞻远瞩。不过由于节能方面的问题以及对构造物理要求的忽视，人们一度丧失了对这种材料的过分乐观。

如今，所幸人们转向探寻合理的节能方案以及使玻璃有效隔热、控制阳光的革新技术，玻璃重新又成了性能良好的建材。它满足了功能与设计的双重要求，也开创了崭新的应用领域。

尽管玻璃给了我们多样的可能，但不应忘记，它是一种很脆弱的材料。如果受到过大的集中应力，玻璃会毫无预兆地突然破碎。这就需要人们精确地掌握玻璃的性能，并精心设计和建造玻璃结构。

本书将循序渐进地介绍给同学们建材玻璃的基本知识和玻璃构造方面的内容。在前三章，读者将学习到当今玻璃的性能与种类，然后是玻璃构造基本的原则，最后是不同的应用范围及局限性。技术原理的解释浅显易懂，按照原理编排并辅以简单实例。这样，同学们可以了解到当今玻璃技术的概况，并能运用玻璃来设计自己的项目，使构想实现。

P11

# 建材玻璃

P11

成分

## 玻璃制造

玻璃是由加热硅土（二氧化硅）和钠化合物（碳酸钠）为主的混合原料而产生的。钠化合物起所谓的“助熔剂”作用，即，将硅土很高的熔点（约1700℃）降低。于是原材料会在超过1100℃时熔化，这种熔化是无定形的——即几乎不形成晶体。由于玻璃的结构类似于液体，玻璃有时也被称作“过冷液体”。〉见表1

玻璃种类

建筑中最常用的玻璃是钠钙硅酸盐玻璃，其主要组分为二氧化硅、氧化钙和氧化钠。将成分中的氧化钙换成氧化硼可以制成硼硅酸玻璃，由于其化学稳定性和热稳定性好，往往被用作防火玻璃。用铅金属及其他原料做成的铅玻璃以及其他一些特种玻璃，如用于光学仪器的玻璃，在建筑领域不常用。而彩釉玻璃则开始愈来愈多地用于幕墙。透明有机玻璃，如丙烯酸玻璃及聚碳酸酯玻璃，比无机玻璃更轻、工作性更佳，但由于表面硬度较低，很不耐刮擦，因此不够耐久。

P11

## 基本产品

经高温制成或在冷却后直接形成的玻璃产品一般称为“普通玻璃”。建筑中会使用各种普通玻璃。除了透明光滑的平板玻璃，还会使用有特殊表面设计或特殊形状的玻璃。普通玻璃通常要进行深加工或修整。〉见“深加工及修整”一节

接下来讲解使用于建筑的普通玻璃及其制造方法。

P12

浮法工艺

### 浮法玻璃

浮法玻璃是最常见的普通玻璃。它由制造工艺得名。浮法生产工艺始自1960年，是平板玻璃生产历史上的一大突破，因为它第一次使得大规模地生产清澈透明的、近乎纯平的玻璃成为可能。

**表1**
**玻璃的组成（依据EN 572，第一部分）**

| 二氧化硅 | $SiO_2$ | 69%～74% |
|---|---|---|
| 氧化钙 | $CaO$ | 5%～12% |
| 氧化钠 | $Na_2O$ | 12%～16% |
| 氧化镁 | $MgO$ | 0%～6% |
| 氧化铝 | $Al_2O_3$ | 0%～3% |

生产中首先是将称为“一批”的原材料放进熔炉里熔化。然后，熔融的玻璃被放到熔化的金属锡上。由于玻璃相对密度较小，会漂浮于锡液之上，由此而形成了平滑的表面。这样就生产出一条连续的玻璃带，玻璃带慢慢变硬；其厚度取决于将它拖过锡液槽的速度。玻璃带在通过锡液槽（又称“浮槽”）之后，要再通过冷却区，最后被切成一片片。〉见图1

标准规格，即所谓“带型”，是600cm×321cm。其标准厚度（即“标厚”）为2mm，3mm，4mm，5mm，6mm，8mm，10mm，12mm，15mm和19mm。

P12

### 平板玻璃或窗玻璃

“窗玻璃”一词有些歧义，因为现在通常是用浮法玻璃来做镶窗的玻璃。而平板玻璃或窗玻璃现在一般只使用玻璃拉拔设备进行少量生产，生产时玻璃带是从熔炉里水平地甚至竖直地拉出来的。如今，这种工艺只用于生产特殊的彩色玻璃或非常薄的特殊玻璃。玻璃的表面比浮法玻璃差一些，有较明显的波纹（称为“线道”）。

P13
辊轧工艺

### 压花玻璃

压花玻璃或称滚花玻璃是用辊轧方法制成，也就是玻璃料从两个水冷的辊子间通过，压成连续的玻璃带。辊子的表面刻有图案，使得玻璃带有了纹理。在辊轧玻璃时再加入丝网，就可以制造出夹丝玻璃或装饰用夹丝玻璃。因为有表面的纹理或装饰，压花玻璃又被称作装饰玻璃，〉见图2用于隔断或立面窗户等不希望也不需要让人一眼望穿

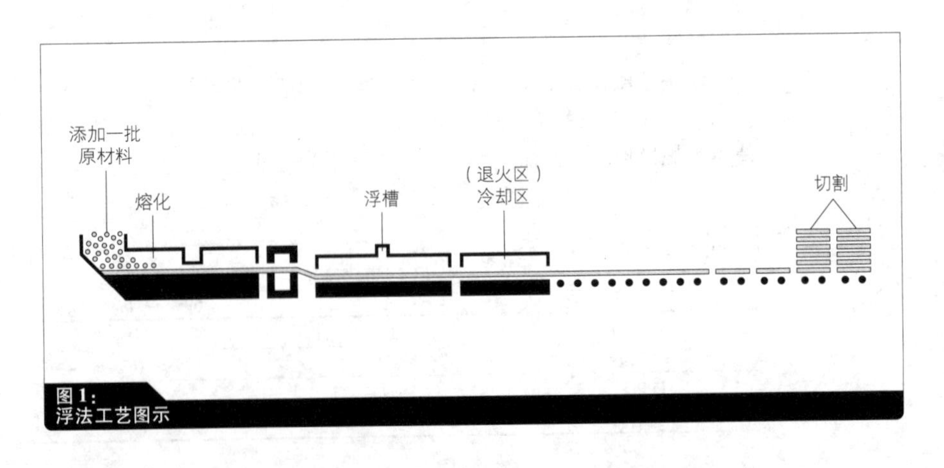


**图1：**
**浮法工艺图示**

图 2:
装饰玻璃

图 3:
不同类型的 U 形玻璃

的地方。〉见“设计玻璃”一章，“装饰玻璃”一节

P13

U 形玻璃

U 形玻璃的生产工艺与压花玻璃类似。这种玻璃不但有表面纹理，它还有特殊的（U 形的）断面形式，因此具有结构优势，可以用于更大的跨度。〉见图 3

由于较为经济，一直以来 U 形玻璃被用于工业建筑。现在，U 形玻璃还在其他种类的建筑形式中得到广泛的应用。〉见“应用”一章，“U 形玻璃”一节

P14
玻璃砖

压熔玻璃制品

压熔玻璃制品是玻璃砖、顶棚玻璃片和玻璃混凝土的通称。这些产品是将两块玻璃熔化经模压成型的（压熔法）。待冷却后玻璃砖中间空腔里的气压较低，因此几乎不会发生冷凝。玻璃砖一般用于室内或作为半透明构件用于建筑外立面实墙。建筑玻璃主要用于钢筋混凝土结构，这是因为它适宜承受较高的荷载。〉见图 4

P14

**深加工及修整**

多数基本玻璃产品生产出来之后都要进行深加工或修整，这样不

图 4：
玻璃砖

但可以调整形式或形状，而且能改善物理性能和结构特性。加工工艺从机械处理或热处理到表面镀膜或表面艺术设计，不一而足。

P14

机械加工

切、钻、磨及抛光之类的工艺通称为机械加工或机械修整。

切

玻璃被切成需要的形状。这个过程并不是真的切，因为切割轮或金刚石只是在玻璃表面刻下划痕，之后沿着划痕轻轻弯曲玻璃使之断开。从浮法生产线上出来的玻璃直接被切成“带型”（600cm×321cm）大小，之后经过修整做成想要的尺寸。切割及其他的深加工都是由机器完成。比如，使用喷水切割机可以精确地做出复杂的形状。切割时除使用切割工具（金刚砂）之外，还要辅以高压喷射水（水压高达600MPa）。

处理边缘

由于切割后玻璃的边缘仍然尖利，需要处理边缘以防伤人，同时便于成品制造。下面的图表将对处理工艺进行了仔细的区分讲解。〉见图5及表2

〉

提示：

边缘处理不仅对玻璃原片的光学性能有所影响，也会影响其稳定性。边缘不均匀、尖利增加了玻璃损坏（破裂、剥落）的可能性。因此，边缘处理的类型必须在签订供货合同前就确定下来，有时还要明确更具体的样式。

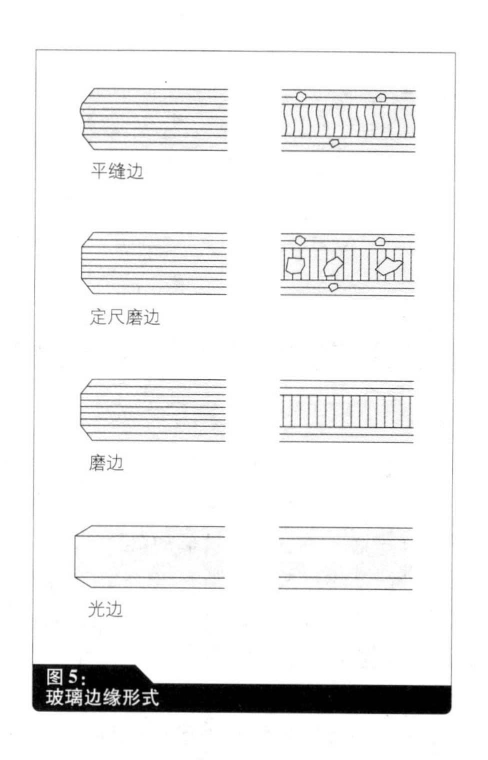


**图 5：**
**玻璃边缘形式**

**表 2**
**边缘处理方式**

| 名词 | 定义 |
| --- | --- |
| 切边 | 切割平板玻璃而成，边缘未经处理呈尖利状 |
| 平缝边 | 切割边缘，并用磨细工具磨平边缘 |
| 定尺磨边 | 磨至需要尺寸的面板玻璃。边缘或有光点及碎屑 |
| 磨边 | 所有边缘被磨成半粗糙表面。边缘不允许出现光点及碎屑 |
| 光边 | 将磨细的边缘抛光 |

开孔

现在，有许多不同应用形式的玻璃构造都需要在面板玻璃上开孔。有了孔，就能在这个位置装紧玻璃片。由于玻璃又硬又脆，进行机械处理必须使用合适的工具——以开孔为例，就需要用一种带金刚石尖端的水冷空心钻孔机。钻孔时为防止破裂，要在玻璃两面同时钻。由于孔内壁的集中应力可能会很大，玻璃板片在钻孔后需要进行热处理，以提高玻璃强度。

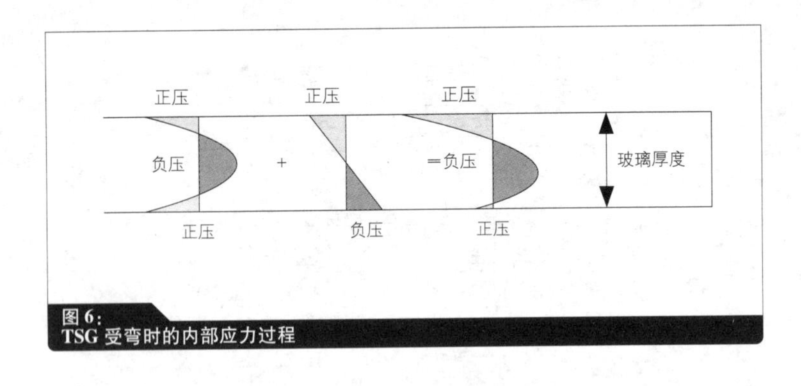

图 6:
TSG 受弯时的内部应力过程

P16

热处理

热弯

将平板玻璃加热到600℃左右使其软化，然后将其弯曲成型就产生出曲面玻璃或弧形玻璃。玻璃的弯曲可以是沿单一轴线（即柱面）的或是沿两个轴线（球面）的，如全玻璃的穹顶。

热处理

钢化玻璃（TSG）是将玻璃入炉在特定条件下加热至约 600℃，然后迅速降温形成的热处理玻璃。玻璃内部因加工产生的应力状态被“冻住”了，由此大大提高了材料的弯曲强度。〉见图 6

另外，钢化玻璃在耐热冲击方面比普通浮法玻璃强很多。钢化玻璃可耐受 150K 的急剧升温，而普通浮法玻璃只能耐受 40K。不过，钢化玻璃被视为安全玻璃的主要原因在于其破碎的方式。由于钢化玻璃内部始终有应力，它破碎时会一下碎成很多边缘不是很尖锐的小碎片，这也就大大降低了造成严重伤害的风险。一分为二地看，好处在于由于碎片仍会连结在一起，整片碎玻璃一般都还留在窗框内；但由于也有些相对大一点的碎片有坠落的可能，因此连结在一起的碎片仍对窗下的人构成威胁。

均热检验

虽然不常发生，但钢化玻璃有一个不利的特性，即可能自爆：肉眼看不见的微小的硫化镍杂质会渐渐膨胀，可能使玻璃突然破碎，这种情况即使在玻璃安装后数年仍可能发生。一种可靠的监测硫化镍杂质的方法是均热检验，玻璃片要加热到约 290℃，在热浸箱中浸泡约 4 小时，含有硫化镍杂质的 TSG 玻璃很可能破碎，这样就提前避免了将其安装到建筑上。根据 EN 1419 的规定，这样均质化处理过的钢化玻璃编号为 TSG-H。

热处理后，钢化玻璃内部会产生方向性（各向异性）的结构形态，在特定的日光或偏振光条件下可以看到。光线在有应力的区域里折射了两次，于是将纹理或团状的结构通过光谱色显现出来。

半钢化玻璃

半钢化玻璃（HSG）的生产过程类似于钢化玻璃。不过它是缓慢冷却的，由此减小了表面压力。因此，半钢化玻璃的弯曲强度小，破碎的模式也不同于钢化玻璃。半钢化玻璃破碎时类似于未经热处理的玻璃：从破碎中心向外辐射出几条裂纹，相应地，碎片也大。由于半钢化玻璃破碎后造成严重扎伤的危险较钢化玻璃大，它不被归为安全玻璃一类。

不同于钢化玻璃，半钢化玻璃没有硫化镍杂质晶体带来的自爆问题。在个别情况下，可以对半钢化玻璃的边缘进行处理（磨），但是一般情况下还是要在热处理之前进行处理。

P18

### 化学钢化处理

将玻璃浸蘸电解液的化学处理可以令玻璃表面产生收缩力。该处理还可用于钢化处理具有复杂空间形态的很薄的玻璃。不过，化学钢化玻璃在建筑中较少见。

P18

### 夹层玻璃和夹层安全玻璃

夹层安全玻璃（LSG）的组成至少要有两层玻璃及中间夹的一层聚乙烯醇缩丁醛（PVB）薄膜。〉见图7，图8

这种玻璃常用于采光顶或其他防止人员坠落的地方，也用于汽车生产。在这些地方使用夹层安全玻璃的一个根本原因，就是它可以将碎片粘住。玻璃面板破裂时，玻璃碎片大多都还粘在透明的 PVB 膜上。薄膜很难被刺穿，于是就大大降低了伤人的风险。

具黏弹性的 PVB 膜对玻璃的黏性好，抗撕裂力强，透明度高并且耐日照。PVB 膜的标称厚度为 0.38mm，如有需要，厚度也可达到此值的数倍。因此，标称厚度有 0.38mm、0.76mm 及 1.52mm 几种。

**提示：**

由于本身有应力，钢化玻璃不能再被切、钻、磨等，否则将破碎。因此，诸如此类所需的机械加工必须在热处理过程之前完成。

**提示：**

需要注意，半钢化玻璃的弯曲强度为 70N/$mm^2$，钢化玻璃的弯曲强度为 120N/$mm^2$，普通玻璃的为 40N/$mm^2$。耐热冲击性方面，普通玻璃可耐受 40K，半钢化玻璃为 100K，钢化玻璃为 150K。以下是估算各种玻璃的抗冲击强度、抗压强度及耐高温的简易计算系数：普通玻璃为 1，半钢化玻璃为 2，钢化玻璃为 3。

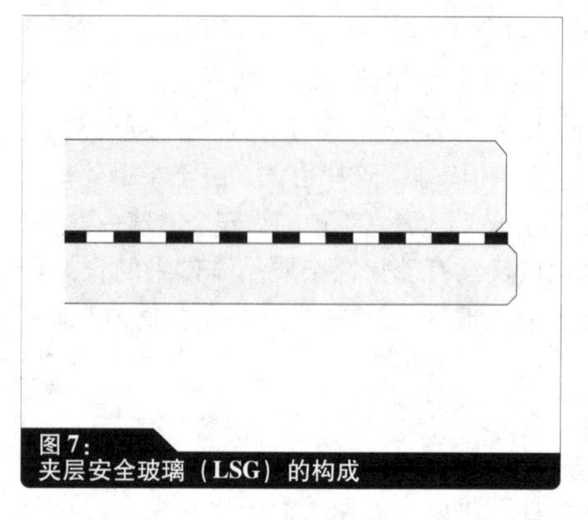

图 7：
夹层安全玻璃（LSG）的构成

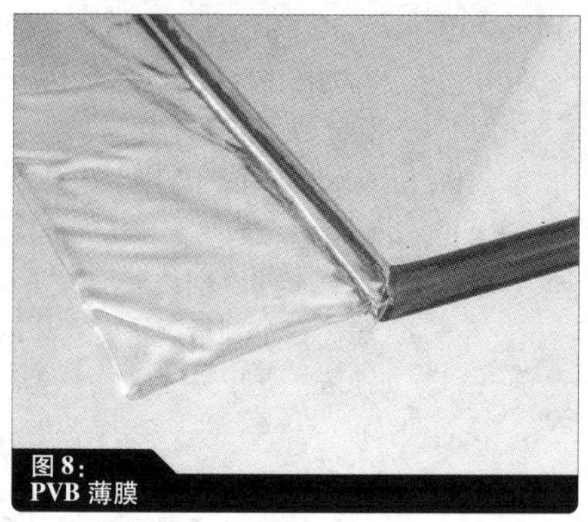

图 8：
PVB 薄膜

生产夹层安全玻璃要经过以下三步：首先，在清洁的空间里准备好要夹在玻璃和薄膜之间的夹胶；然后，玻璃和薄膜要通过初步夹胶室，在这里辊子的温度和压力使玻璃和薄膜的粘结力加强；最后的胶结是在高压高温的高压釜中完成的。完成了这些步骤，原来模糊的薄膜才会变得非常透明。

用其他材料作中间夹层的多层玻璃统称为夹层玻璃。这些玻璃一般达不到安全玻璃的要求，因此除非有特定的验证，这些玻璃不归为安全玻璃之列。若生产有光伏组件的组合夹层玻璃，则要使用 EVA 薄膜。

P19

### 多层中空玻璃

为提高建筑中窗户的隔热性能，现在中空玻璃几乎是不二的选择。将两片或多片玻璃沿着周边粘结，即为多层中空玻璃（MIG）。玻璃之间的空气间层里是密封的干燥空气或惰性气体。

封边

绕面板玻璃一周的封边由铝、不锈钢或塑料的间隔框及密封胶组成。〉见图 9

间隔框里填有除湿剂，可以吸收气体或空气中的水分残留。这样有助于防止空气间层里发生冷凝。密封包含两道密封胶。第一层为丁基密封胶，它同时还能将间隔框粘在玻璃上；第二层为耐候弹性密封胶，如聚硫、聚氨酯或硅胶等。

目前，玻璃之间的气体间层里一般充氩、氪，或比较少见的如氙等惰性气体，这是因为惰性气体比干燥空气更能够提高玻璃的隔热性。不过，能显著提高中空玻璃隔热性能的还是玻璃表面的镀膜。

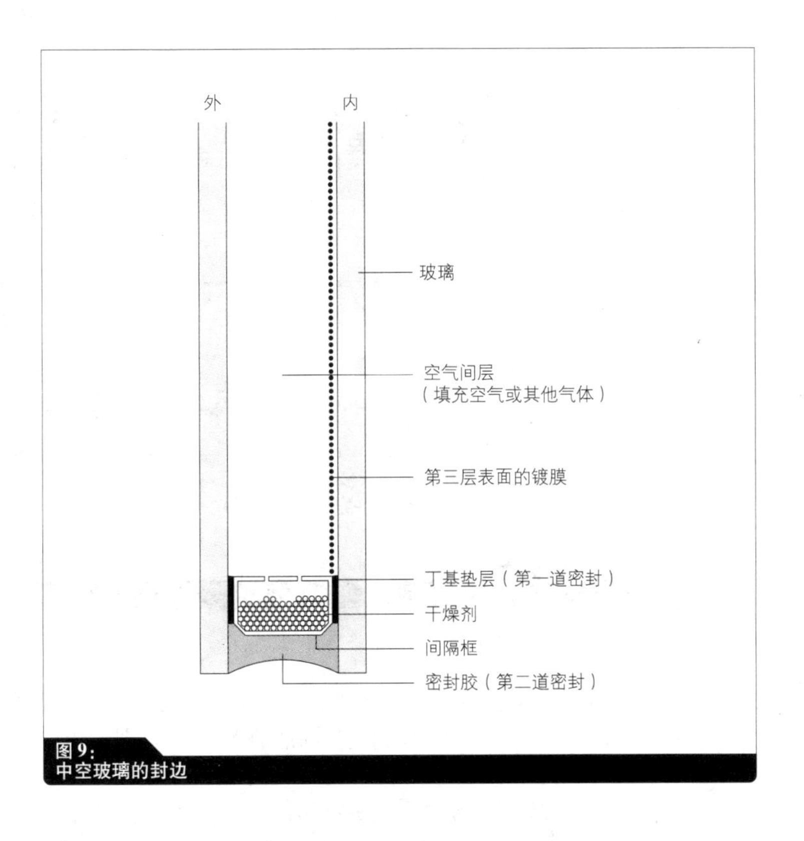


**图 9：**
**中空玻璃的封边**

“中空玻璃效应”

由于中空玻璃的空气间层是密闭的，内部截留气体和大气之间的压力差会使玻璃片产生内凹或外凸。〉见图 10

这会造成玻璃片的反射扭曲。压力差造成玻璃片的额外应力，称气候应力。玻璃受到拉力，特别是在边缘密封处。在中空玻璃的尺寸较小或很窄的情况下，这一效应更为突出，还可能造成边缘密封受到过应力，并过早失效。

P21

### 表面镀膜

根据需要，可将玻璃表面镀膜，以显著改善玻璃的光学性能和物理性能。一般来说，镀膜材料是金属或金属氧化物。目前，可以生产多种镀膜玻璃产品，其中有隔热型或阳光控制型的，有反射型或无反射型的，有不同颜色的，还有自洁型的。薄薄一层镀膜不会影响玻璃的结构特性，不过膜本身一般无法耐受环境影响（腐蚀）或机械作

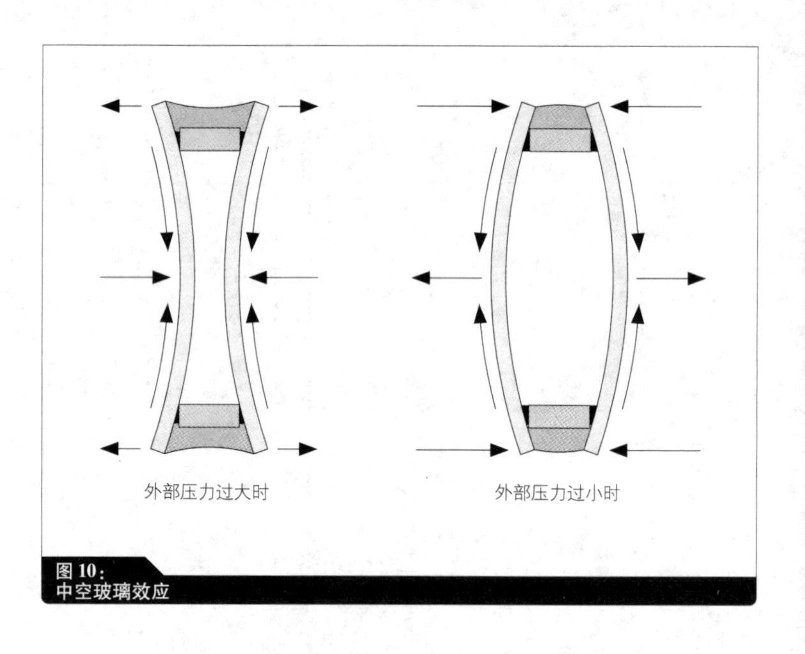

图 10：
中空玻璃效应

用（刮擦）。因而，很多镀膜，特别是具有显著隔热或阳光控制性能的涂层，只能用于朝向空气间层的表面。（即图 11 中的第二、第三层表面）。〉见图 11

镀膜可在生产浮法玻璃的过程中（“在线式”，过去的做法）或生产之后（即“离线式”）进行。在线式工艺是趁着玻璃还热时将金属氧化物液体喷涂在玻璃表面，高温使涂层紧紧附着于玻璃（热解）。这样做出来的膜（称“硬膜”）很坚固，可用于玻璃的外表面（图 11 中的第一层表面）。

如今，多数隔热或遮阳的玻璃都是由先进的阴极射线法（真空磁控溅射）生产的，用这种方式可以涂覆多层极薄（厚度以纳米计）的膜。也可以使用溶胶工艺镀膜，这种工艺是将玻璃多次浸蘸化学溶液。每次浸蘸后要烘烤玻璃表面。

P22

表面设计

除了表面镀膜，还有各种装饰玻璃表面的工艺。生产客户定制产品的一个常用方法，就是在整个或局部的玻璃表面上印刷。除了印刷，酸刻玻璃和毛玻璃也具有半透明的表面。〉见“设计玻璃”一章

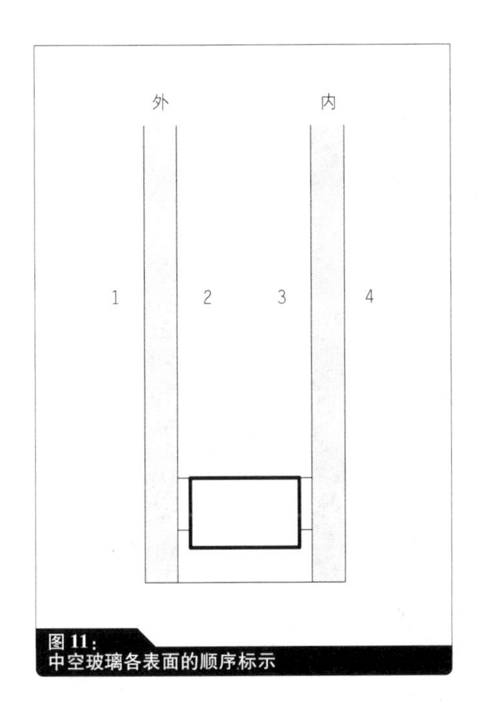

**图 11：**
**中空玻璃各表面的顺序标示**

P23

## 特殊用途玻璃

P23

### 中空隔热玻璃

在寒冷或温和气候地区，由于节能的要求，使用中空玻璃自然顺理成章。在 1980 年代，玻璃窗还是造成高昂年度取暖费用的主要原因。而如今，即使是使用了大面积玻璃的建筑，其化石能源消耗也可以很少，甚至为零。

传热系数

根据 EN 763，表征建筑材料（包括窗及玻璃）的热损失性能的物理量为传热系数（$U$ 值）。该值代表当室内外的温度差为 1K 时，每 1 平方米建筑材料上通过的热流（单位 $W/m^2 \cdot K$）（正确的概念应该限定是“1 小时内通过的热流”——译者注）。不过，影响窗户隔热性能的最关键因素是玻璃与窗框的构造。因此，玻璃的 $U$ 值（$U_g$）与整个窗户的 $U$ 值（$U_w$）是不同的。因为在窗户角部的热量传递一般高于窗户的中间部位，与 $U_g$ 值相比，$U_w$ 值较大，即性能较差。$U_w$ 的计算如下：

$$U_w = \frac{U_g \times A_g + U_f \times A_f + \varphi \times L_g}{A_w} \quad (W/m^2 \cdot K)$$

$U_w$ 窗户的传热系数

$U_g$ 面板玻璃中央处的传热系数

$A_g$ 玻璃面积

$U_f$ 窗框的传热系数

$A_f$ 窗框面积

$\varphi$ 玻璃边缘导热性能的线性系数

$L_g$ 玻璃边缘的长度

$A_w$ 整个窗的面积

两层无镀膜玻璃是以空气层间隔的中空玻璃，其 $U_g$ 值大约为2.8～3.0W/m² · K。比较而言，如果以惰性气体填充并镀膜（一般镀于第三层表面），充氩气时 $U_g$ 值约 1.1W/m² · K，而填充氪气时约为1.0W/m² · K。3层玻璃的中空玻璃填充氩气时，$U_g$ 值约为0.6W/m² · K，填氪气时约为0.5W/m² · K。〉见表3

就中空玻璃而言，其三分之一的热传递由对流和传导造成，而另外三分之二是由于辐射。在气体介质中的能量传递称为“对流”。由于两片玻璃之间有温度差，在空气间层中的气体就会运动，由此将热从高温处传到低温处。热传导是能量从固体——就当前问题而言，是能量从玻璃——向其对边的传递。玻璃板片间的热辐射则是有温度差的玻璃表面之间的直接辐射换热。〉见图12

**表3：**
**不同种类中空玻璃的特性**

| 构造组成（mm） | $U_g$ 值（W/m² · K） | $G$ 值（%） | $L_t$ 值（%） |
|---|---|---|---|
| 双层中空玻璃 4/16/4，氩气 | 1.1 | 63 | 80 |
| 双层中空玻璃 4/10/4，氪气 | 1.0 | 60 | 80 |
| 三层中空玻璃 4/14/4/14/4，氩气 | 0.6 | 50 | 71 |
| 双层中空玻璃 4/12/4/12/4，氪气 | 0.5 | 55 | 72 |

$U_g$ 值：玻璃的传热系数

$G$ 值：太阳辐射得热系数

$L_t$ 值：透光率

此表数值根据厂商具体产品得出，不代表所有情况。

低辐射玻璃

热保护镀膜起到降低热辐射造成的能量损失的作用，因此这层镀膜又被称为Low-E（即Low-Emissivity，低发射率）膜。银镀膜现在最常用，因为其兼具了极低的发射率与中性色调、高透率的优点。隔热中空玻璃与无镀膜的中空玻璃用肉眼几乎无法区分。〉见图13

冷凝

尽管中空玻璃有很好的隔热性能，但当室外温度很低时，仍会在其内表面的边缘发生冷凝。随着时间推移，冷凝水会损害窗边密封，对木窗则会损害压玻璃条。由于间隔框（一般为金属材质）形成了热桥，导致边缘处有较高的热传导而造成冷凝。使用不锈钢或者塑料取代铝作为间隔框的材料，可以降低玻璃边缘的导热性能的线性系数（$f$），由此可减少冷凝并在几个百分点的范围内改善窗户的 $U_w$ 值，具体改善程度根据尺寸而定。

太阳辐射得热系数

除了 $U$ 值以外，中空玻璃还有一个重要的特征值，那就是太阳辐射得热系数（$G$ 值）；$G$ 值（依据 EN 410）表征了透过玻璃进入的太阳辐射能量。$G$ 值为初次透射的太阳辐射及玻璃吸收的太阳辐射部分（二次透射）之和，包括热辐射和对流形式的能量。对被动式太阳房而言，$G$ 值高有利于被动获取太阳能量，因此有益。对其他类型的建筑，比如开窗面积大的办公楼，吸收太多太阳辐射则会带来内部过热的问题。〉见“阳光控制玻璃”一节

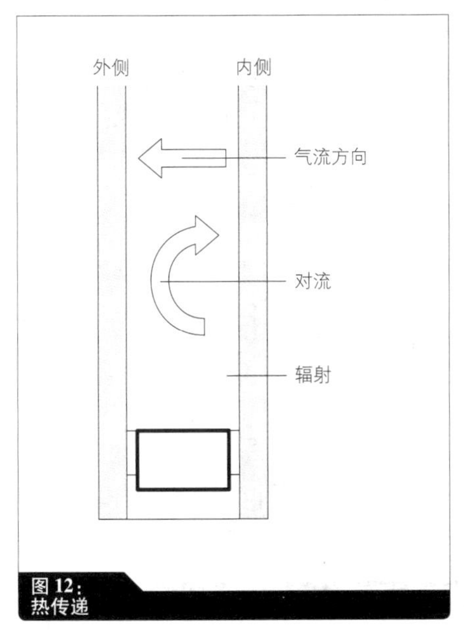

图12：
热传递

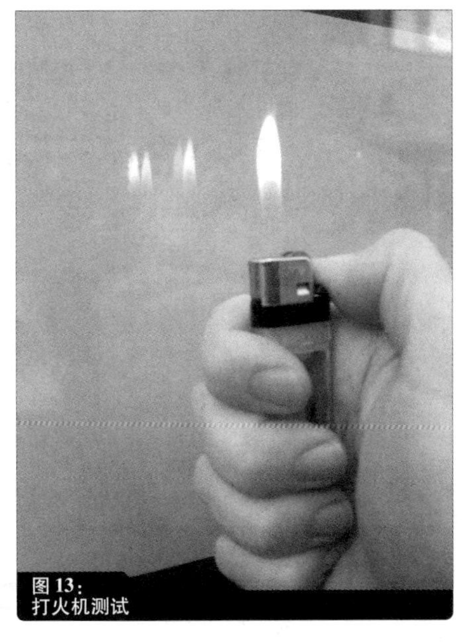
图13：
打火机测试

**注释：**

在中空玻璃安装之后，可以根据打火机测试确定镀膜是在玻璃的哪层表面。每片玻璃反射火苗两次。火苗在有镀膜表面的映像与在其他表面的映像不同。

**提示：**

改变中空玻璃的镀膜位置会改变 $G$ 值。双层中空玻璃的镀膜如果在第二层表面会比在第三层的 $G$ 值小。3 层玻璃的中空玻璃当镀膜在第三、第五层时比在第二、第五层时的 $G$ 值更高。不过这种改变不影响 $U$ 值。

中空隔热玻璃的 $G$ 值大致介于 0.6 ~ 0.65 之间。3 层中空玻璃的 $G$ 值还会更低些。〉见图 14

P26
温室效应

**阳光控制玻璃**

作为透明的建筑材料，玻璃可以透过短波（波长为 300 ~ 3000nm）太阳辐射，而长波（ > 3000nm）的热辐射则无法透过。很多进入室内的太阳辐射被照到的表面吸收，转换为热，并再次以长波的形式辐射。这就无法通过玻璃再回传到室外了，因此室内会持续升温。这种温室效应意味着即使室外气温较低，玻璃房内仍有过热的可能。〉见图 15

阳光控制玻璃会吸收和反射照到玻璃上的辐射，以此将很多辐射能阻挡于室外。阳光控制玻璃多使用本体着色玻璃；它吸收一部分辐射，不利的是它也吸收了可见光。最早的镀膜阳光控制玻璃则有可见光反射量过大的缺点。

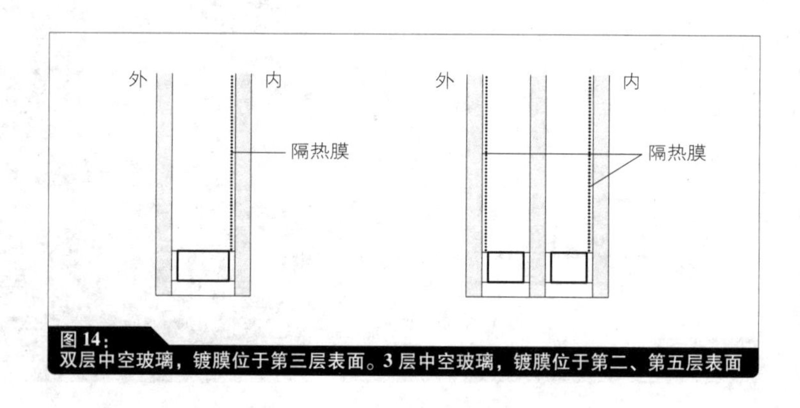


**图 14：**
**双层中空玻璃，镀膜位于第三层表面。3 层中空玻璃，镀膜位于第二、第五层表面**

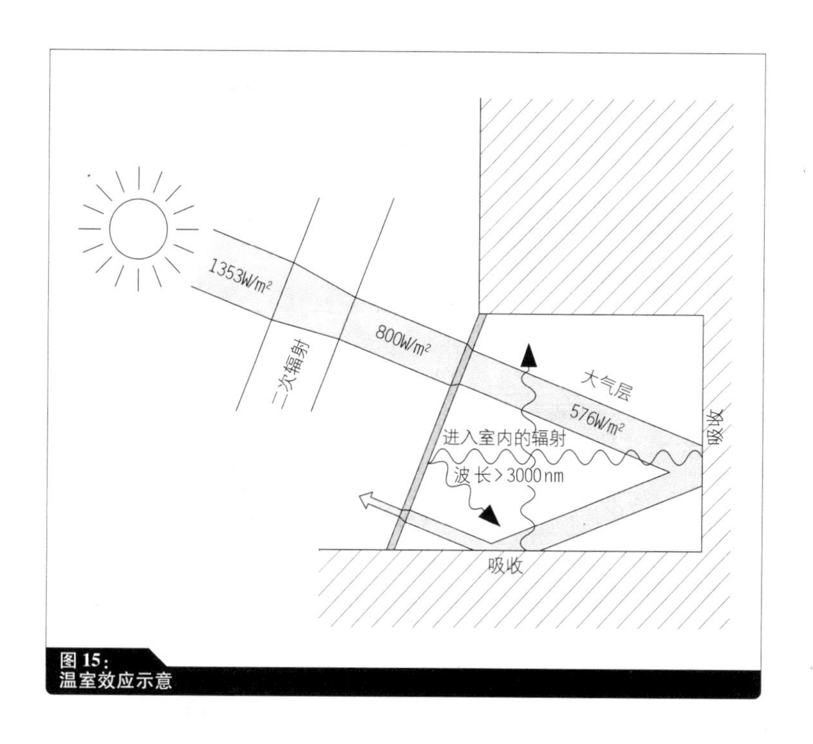


**图 15：**
**温室效应示意**

选择性

现在的阳光控制型玻璃具有选择性镀膜：即，它对可见光为透明，但会反射或吸收掉长波的红外辐射；其 $G$ 值目前介于 20% ~ 50% 之间。

为获取更高品质的室内天然采光，在选用阳光控制型玻璃时还要挑选高选择性（$S$）的品种。选择性是透光率对 $G$ 值的比值。如，窗的 $G$ 值为 40%，其可见光的透过率为 76%，则其选择性为 76∶40 的比值，即 1.9。该特性的理论限值为 2.0。

显色性

不仅仅是进光量对室内天然采光有重要意义，显色性也很重要。显色性指数应不小于 90%，这是衡量天然光显色性能的标准，通过在室内测量反射阳光的表面得出。玻璃的显色性指数（$R_a$）最高可达 99%。〉见表 4

折减系数 $F_c$

很多情况下，可以通过内外遮阳（如百叶、百叶窗帘、雨篷等）来改善阳光控制的效果。计入窗遮阳的太阳辐射得热系数称为 $g_{total}$，是玻璃的 $G$ 值与一个折减系数的乘积。〉见表 5

$$F_c \ (g_{total} = F_c \times G)$$

**表 4：**
**不同种类中空玻璃的特性**

| 构造组成 | $U_g$ 值（$W/m^2 \cdot k$） | $G$ 值（%） | $L_t$ 值（%） | $R_a$ |
|---|---|---|---|---|
| 单玻，6mm 厚 | | | | |
| 无色 | 5.7 | 56 | 45 | — |
| 绿色 | 5.7 | 45 | 53 | — |
| 双层中空玻璃<br>充氩气<br>标称值 68/34 | 1.1 | 36 | 66 | — |
| 双层隔热玻璃<br>充氩气<br>标称值 40/21 | 1.1 | 22 | 40 | 88 |
| 双层隔热玻璃<br>6+16+4，充氩气<br>标称值 50/27 | 1.1 | 29 | 50 | 95 |

$U_g$ 值：玻璃的传热系数

$G$ 值：太阳辐射得热系数

$L_t$ 值：透光率

$R_a$：显色性指数

此表数值根据厂商具体产品得出，不代表所有情况

**表 5：**
**折减系数 $F_c$**

| 条目 | 遮阳的类型 | $F_c$ |
|---|---|---|
| 1 | 无遮阳 | 1.0 |
| 2 | 内遮阳或中空玻璃间层内置遮阳 | |
| 2.1 | 透明度低或白色的反射表面 | 0.75 |
| 2.2 | 浅色或较透明 | 0.8 |
| 2.3 | 深色或较不透明 | 0.9 |
| 3 | 外遮阳 | |
| 3.1 | 旋转百叶，背后通风 | 0.25 |
| 3.2 | 百叶窗帘及低透明度的织物，背后通风 | 0.25 |
| 3.3 | 普通百叶窗帘 | 0.4 |
| 3.4 | 卷帘、百叶挡板 | 0.3 |
| 3.5 | 凉棚、外廊、悬空百叶片 | 0.5 |
| 3.6 | 雨篷，有顶部及侧面通风 | 0.4 |
| 3.7 | 普通雨篷 | 0.5 |

**注释：**

阳光控制玻璃（包括所谓无色的阳光控制玻璃），由于成分不同，外观的反射程度和色调均不同（如，呈蓝色、绿色或银灰色）。即使已知其特征值，在安装之前仍需取样测试其成分。这在替换单块玻璃时尤为重要。

在面板玻璃的外面（局部）做彩釉印刷也可降低 *G* 值。印刷的密度或范围越大，*G* 值越小。〉见“设计玻璃”一章，“彩色玻璃”一节

P29

**隔声玻璃**

建筑项目一般需要达到最低隔声标准。根据功能用途决定隔声程度要求。通常需要区分空气噪声和撞击噪声两大类。撞击噪声是通过在建筑局部行走或重击传递出来的。而通过空气传递的噪声，如谈话声或交通噪声，称为空气噪声。

声压级是衡量噪声的物理量，单位为分贝（dB）。建筑立面的有效隔声量主要取决于其窗户的隔声性能，即，其窗户玻璃及框架的结构。一般的中空玻璃的隔声性能比普通玻璃显著优越。以下方法可以进一步改善双层中空玻璃的隔声指数：

— 增加玻璃的层数并采用非对称的构成（玻璃的厚度不同）；

— 增厚玻璃之间的空气间层；

— 使用夹层安全玻璃；

— 使用有特殊夹层或铸塑树脂的夹层玻璃或夹层安全玻璃（LSG）。

隔声膜要达到与传统 PVB 膜相同的安全性（抗拉强度和防碎裂性能）要求，因此极少使用铸塑树脂夹层隔声玻璃。铸塑树脂的抗拉强度较低，并且使用一段时间后会松脱及移动，露出夹层玻璃的边缘外。〉见图 16

P30

**防火玻璃**

防火玻璃用于设计为阻燃或耐火的墙面或立面窗。应该特别注意需要的玻璃防火等级，等级划分根据 EN 13501 – 1。〉见表 6

**提示：**

计权隔声指数（$R_w$）是量度中空玻璃隔声量的一个物理量，要由建筑法规中认可的权威机构测定并有证书证明。其数值是在实验室中计算，其中没有计入通过建筑相邻部分的传声。与之相比，计权隔声指数（$R'_w$）（此处原文有误，$Rw_w$ 与 $R'w_w$ 的英文名词完全相同。因为未查到 $R'_w$ 的准确定义，需要原作者更正。——译者注）则计入了相邻连接部分的传声量。因此 $R'_w$ 值较 $R_w$ 小。

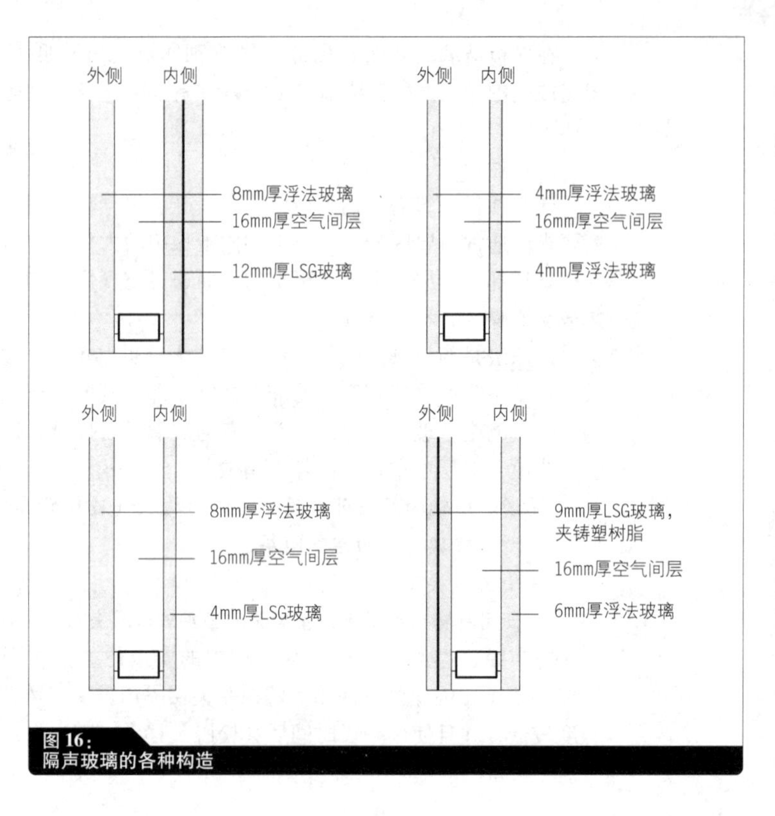

**图 16：**
**隔声玻璃的各种构造**

**表 6：**
**防火玻璃的等级**

| 建筑规范编号 | 在 EN 13501－1 中的防火性能等级 |
|---|---|
| 阻燃 | EI30 |
| | E30 |
| 耐火 | EI90 |
| | E90 |

表中标明的数字表示玻璃的检定防火等级。它是玻璃仍具有至少减缓火灾烟气扩散的能力的时间，以分钟表示（30 或 90 分钟）。E 级玻璃只阻断高温烟气的扩散，而 EI 级玻璃还阻碍火焰带来的高温辐射的穿透。因此，E 级玻璃只能用于火灾时与人有足够安全距离的地方，如距地 1.8 米以上的天窗，或者与疏散通道不相邻的墙壁等。E 级玻璃由单片玻璃构成，而 EI 级玻璃则是由普通平板玻璃与遇高温发泡的中间构造层组成，发泡材料可以在规定时间内阻挡火焰和高温的渗透。

提示：

为确保防火安全性能，所有连接和密封耐火玻璃的材料必须具备认证，并应以适当方式构成框架。

P32

**特殊种类防暴玻璃**

当今，人们对玻璃还提出了抵御恶意破坏、偷窃及暴力犯罪之类的要求。欧洲的标准是根据玻璃所能抵抗的暴力行为的性质和强度来划分特殊玻璃的种类。

防撞击玻璃

防撞击玻璃可以抵挡石头和小型投掷物。这种玻璃由坠球实验来检测。一个重约4公斤的钢球从特定高度接连三次掉落在玻璃上，要求玻璃不被击穿。

防盗玻璃

防盗玻璃能够在特定时间内阻止斧头在上面砍出40cm×40cm的洞口。使用连接在机器上的一把长柄斧做测试。

防子弹玻璃

防子弹玻璃，俗称防弹玻璃，是为抵御各种枪弹——从滑膛枪到步枪——的攻击而造。通过在实验射程试射几种普通枪械来测试它。

防爆玻璃

防爆玻璃用于抵御外部的爆炸袭击。这种玻璃通过人工施加垂直于玻璃表面的压力波来测试。

特殊种类防暴玻璃都是多层结构的，从传统的夹层安全玻璃（LSG）到由普通玻璃、热处理玻璃、薄膜甚至塑料镀膜组成的多层结构的不同种类，其关注点始终是要保护好玻璃后面的人或物。玻璃本身受到严重破坏通常会损坏，并需要更换。防暴玻璃也可装上电网；玻璃破损时电流中断，由此触发警报。电网可以是嵌在夹胶安全玻璃中的银丝网，也可以是喷涂于钢化玻璃一角的导电瓷釉。因为钢化玻璃都会碎裂为很多小碎片，确保了电路能被切断。

P33

**内置构件的中空玻璃产品**

将遮阳百叶或金属网等功能构件内置于中空玻璃空气间层的做法是当前的一种趋势，其优点在于可以保护以下提及的内置构件免于风及其他气候因素的损坏，并避免积尘。

可调式系统

可调式遮阳百叶是内置凹形或凸形的百叶片，其表面有无光的、反光的或穿孔的几种。使用电脑调节可以遮蔽阳光及工作场所的眩光。

可调式遮阳帘也可置入空气间层中。这种构件由织物或穿孔塑料薄片制成。

图 17：
固定式遮阳百叶

图 18：
遮阳格栅

固定式系统

固定式遮阳百叶是预装于空气间层的具有高反射率的横向百叶，百叶片有特殊的截面形状，因此会向外反射出大部分的直射阳光，而将漫射光引入室内。〉见图 17

遮阳格栅的功能近似于固定式遮阳百叶，只不过百叶片纵横交叉。这种复杂的格栅其表面涂覆纯铝以提高反射率，研发这种产品的目的主要为了用于坡度较缓的玻璃屋顶及玻璃穹顶。〉见图 18

内置的折光片或丙烯酸有机玻璃型件的功能也是反射出直射阳光并引入漫射光线到室内。

图 19：
金属网夹层

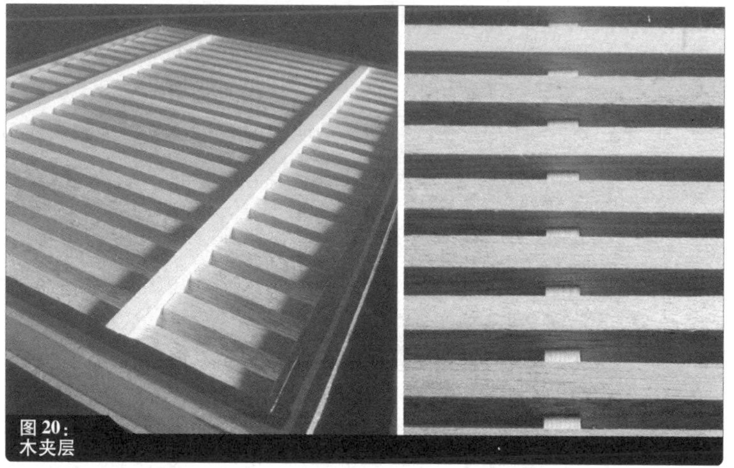
图 20：
木夹层

另外，还可夹入多种不同的金属网夹层。这些夹层不仅可以控制阳光，而且有强化设计效果的作用。〉见图 19

装饰性木夹层由矩形截面的木条构成，可以替代金属网用在中空玻璃间层内。〉见图 20

最后需要提到的是其他一些漫射光线的内置构件，这些构件如果厚度足够，也可起到隔热的作用。中空玻璃内置漫射型微孔薄片由一些紫外稳定性好的微小的透明聚碳酸酯管组成，这些细管会均匀地漫射阳光。在避免直射光照射的地方，如博物馆、画室、体育馆等地方，有时使用这种材料以获取室内天然采光。〉见图 21

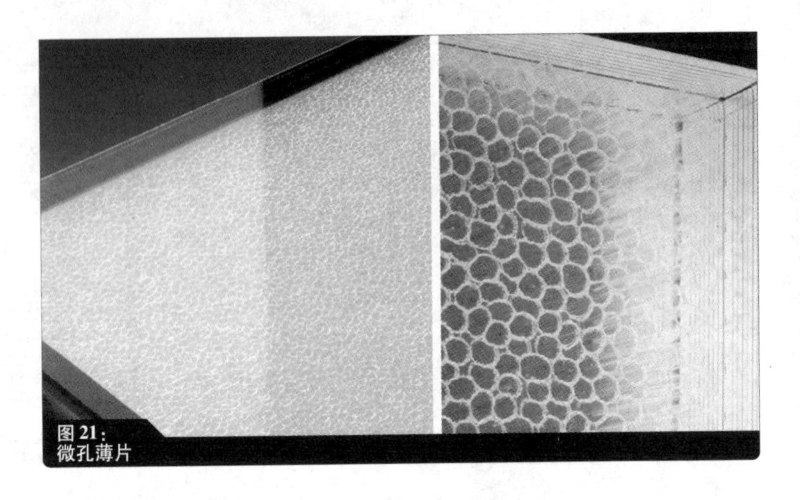
图 21:
微孔薄片

透明保温材料

“透明保温材料”一词有些误导人，因为这种保温材料一般都不是透明的，只是漫射型半透光的。透明保温材料经常被内置于紧贴重质外墙外面的中空玻璃，这样做可以吸收更多的冬季太阳辐射热能，使热能持续传入室内。透明保温材料还可用于嵌入建筑立面，以利用更多的阳光。可以填入两片玻璃之间的材料有塑料微孔薄片、薄壁玻璃管或气凝胶（泡沫颗粒）等。不过，需要对透明保温材料进行遮阳，以防止温暖季节出现室内过热的情况。

P37

**特殊功能层**

玻璃透明、坚硬并耐候，它对多种功能构造层来说都是理想的基层材料。

光伏组件

光伏组件就是一个典型的特殊功能层的例子，这种构件用于建筑立面或屋顶，以将阳光转化成电能。这种构件为两层玻璃片之间夹硅晶电池，或者是直接涂在玻璃上的“薄膜电池”。〉见图 22

可调式功能层

另外，还有一些功能层利用了玻璃对光线和辐射的透射性能，可以根据当时的日照或气候条件进行调节。

提示：

内置构件有一个弊端，它们会增加空气间层的吸热，由此可能导致内置构件的提前老化。没过几年就要更换这些构件将会产生高昂的费用。另外，还要保证新换的材料与原先的外观完全统一。对内置百叶帘，还可能出现因玻璃片的“中空玻璃效应”（因气候应力造成的内凹）导致百叶片卷成筒状而丧失功能。使用厚一些的玻璃可以减轻玻璃弯曲。

图 22：
光伏组件

热致变色层根据温度改变玻璃透光的程度。比如，根据特定的设定，在室温下透明的玻璃会在较高的温度下变成白色，然后反射出去更多的漫射光。

电调光层可通过通电按钮调节，在漫射型半透明或透明的两种不同状态间转换。〉见图 23

电致变色层具有连续调节中空玻璃接收光量及能量的作用。比如，接通电源时面板玻璃会变为深蓝色，减少了透射的阳光及太阳辐射热能。

图 23：
电调光层

图 24：
定制的装饰玻璃

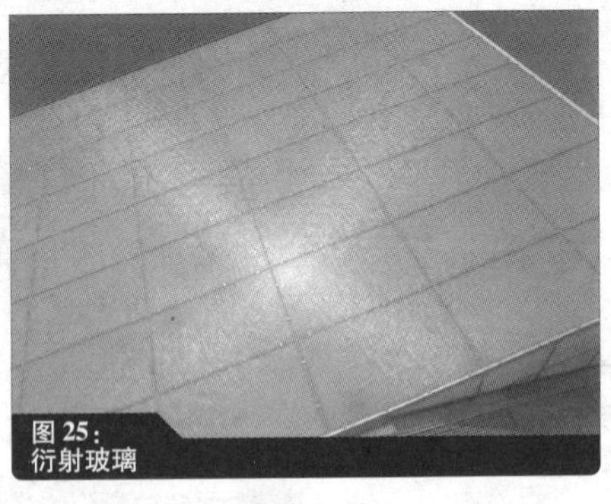

图 25：
衍射玻璃

P38 ## 设计玻璃

P38 ### 装饰玻璃

“装饰玻璃”是指出于功能或设计的考虑而做出表面纹理的压花玻璃。表面纹理的种类从几何图案（方形、矩形、条纹、圆点）到不规则图案，不一而足。如果生产的数量多到可以降低成本，也能按照定制图案生产。〉见图 24，图 25

装饰玻璃的漫射程度可高可低，用于需要漫射光、或在功能或设计方面要避免清晰影像的情况。装饰玻璃可以是单面或双面压花的。其中有一个特殊种类，其表面的凸起图案（如棱镜玻璃）不是用于漫射光线，而是将光线根据入射角度反射或折射向一个特定方向。〉见图 26，图 27

P39 ### 表面磨毛玻璃

毛面玻璃是深受喜爱的建筑设计元素，这种情形合情合理。毛面使得光线的变化更有意思，它还增强了玻璃的实体感。

**提示：**

大多数的装饰玻璃都是热处理玻璃，很多也可以用于制造夹层安全玻璃或中空玻璃。当用于外墙时，应要求厂商证明玻璃的适用性。

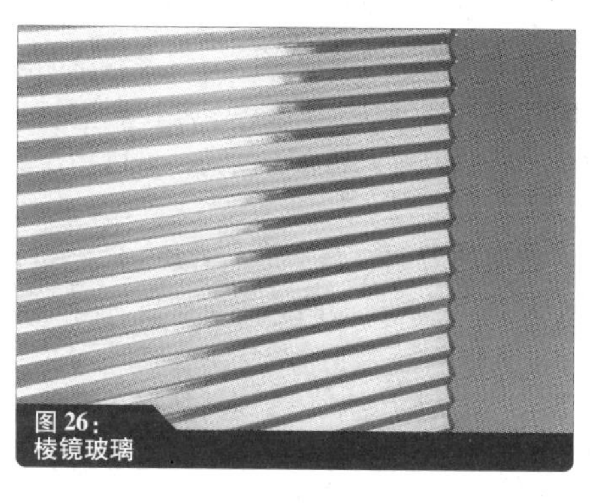

图 26：
棱镜玻璃

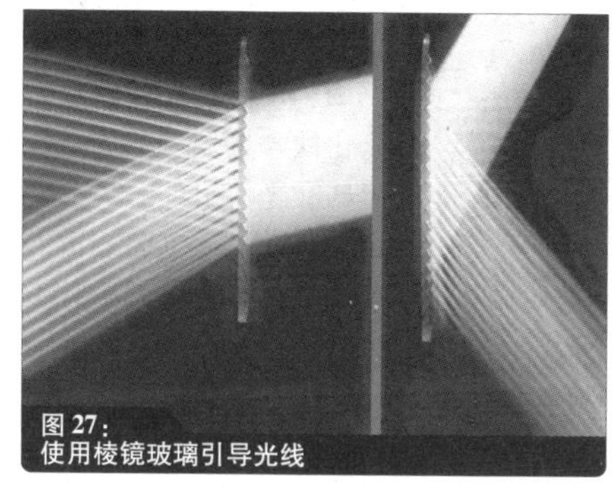

图 27：
使用棱镜玻璃引导光线

蚀刻

蚀刻玻璃是用氢氟酸进行毛面处理的玻璃。在这一过程中，玻璃表层只受到轻微的损伤，因此玻璃的强度损失很小。现在所采用的酸液浓度很低，酸液停留在玻璃表面上蚀刻的时间更长，因此达到的效果还是一样的。处理的时间决定了毛糙的程度。表面越粗糙，玻璃的透明度越低，因为粗糙的表面增加了透射光的漫射。用模板可以蚀刻出各种不同的设计样式、标识或图案。根据厂商的咨询意见，还可以对蚀刻玻璃做热处理甚至做成曲面的。

喷砂

喷砂玻璃的表面是经喷砂粗糙的。由于毛糙处理损伤了表面，玻璃强度降低。经过一段时间，粗糙的表面可能会因诸如油渍残留之类的原因而变色，视觉效果上与蚀刻玻璃近似，可以做图像或图案的效果。

使用丝网印刷的毛面处理是一种不需要将表面变毛糙的方式。方法是涂上半透明的釉料并烧制，使之耐久。

P39

**彩色玻璃**

彩色玻璃近来在建筑界重新复兴。早在中世纪，哥特式大教堂里一片片色彩缤纷的花格窗就曾带给室内美妙的光线变化。现在，有了多种技术可以生产各种不同视觉效果的彩色玻璃。根据成品的不同，“染色步骤”可以在生产玻璃时或在后期处理时进行。

本体着色玻璃

本体着色玻璃的生产是直接将添加剂（金属氧化物）加入玻璃原浆。这种方法可以用于生产彩色浮法玻璃、平板玻璃、压花玻璃以及玻璃砖。不过用于浮法玻璃的色彩范围局限于蓝色、绿色、古铜色和灰色几种，而上面提及的其他玻璃品种则有更多的色彩选择。由于玻璃原浆中有天然的氧化铁成分，无色浮法玻璃实际上带有轻微的绿色调——当玻璃较厚或做毛面处理时这种效果更突出。超白玻璃是指氧化铁含量低、不偏绿色的浮法玻璃。由于其透明度高，可以用于集光器和光伏组件。

熔合

熔合是将不同颜色玻璃浆做成的玻璃结合形成一片玻璃的工艺。〉见图28。不同颜色和形状的玻璃片组合成较大的一片，然后进窑烧至1500℃。这种技术可用于生产平板玻璃，不可生产浮法玻璃。

双色玻璃

滤光玻璃或双色玻璃是使用溶胶方法将厚度不同的几层金属氧化物膜镀到玻璃上。各层薄膜产生相互的干涉使玻璃在不同角度入射光下产生不同的色彩效果。比如使用双色玻璃的立面会将不同入射角度的阳光反射为钴蓝色或金色。滤光玻璃吸收率不高，即它因太阳辐射升温的程度较染色玻璃之类产品为低。

彩釉玻璃

彩釉玻璃和丝网印刷玻璃是两种彩色镀膜玻璃，其工艺是在制造钢化或夹层安全玻璃的过程中（温度超过600℃）将彩色釉层烤到玻璃表面的。因此这两种玻璃多为热处理玻璃；只有使用双组分有机颜料时才可能生产非热处理型产品，而这种产品还不耐刮擦。使用氧化铁含量低的超白玻璃可以获得最好色彩品质。区分这三种不同的连续施釉工艺很重要：

— 在辊轧工艺中，釉彩通过一个有槽的橡胶辊子印到玻璃平板上。
— 在浸涂工艺中，玻璃从浸涂机中通过时颜料被涂在玻璃表面。该工艺与其他工艺不同，是必须使用溶剂的，因此它不是环境友好型的工艺，已经过时了。

（上文提到有“三种”工艺要区分，但下面实际只介绍了两种，似乎原文有误。——译者注）

**提示：**

本体着色玻璃会因太阳辐射升温，因此用于立面时一般先做热处理。熔合彩色玻璃则不能进行热处理，因为不同色彩玻璃的接缝处呈不规则形态。

**提示：**

双色玻璃最大的规格接近1.7m×3.8m。可做热处理，但有严格的限制。镀膜要耐久且耐刮擦，但镀膜不可暴露于室外气候。

图 28：
采用熔合玻璃的彩色玻璃窗

丝网印刷

最均匀的施釉方式是使用丝网印刷。它是在上色台上将颜料挤过精细网线版，印到玻璃表面。可使用的标准色的范围很广。其中有透明、不透明、半透明的各种颜色，还有定制颜色，可以采用连续或间断的方式印到玻璃上。用装饰图案和版式可以将玻璃设计成各不相同的样子。使用电脑制版成像技术（CTS）可以将数字化设计或照片转化为印版。印刷多色图形需要相应数量的印版和工序。这就是说，比如计划要在玻璃上以丝网印刷出四色的照片图像，那就需要四个不同的印版和步骤。〉见图 29，图 30

数字印刷

近来，借助数字印刷技术实现了玻璃上的陶瓷印刷。这种工艺的优点在于图形数据是直接传到印刷机里的，这就不再需要制作昂贵的印版了。而且，它可以同时印刷多种色彩。该工艺特别适合制作复杂的定制图形。

夹层彩色玻璃

现在建筑经常使用夹层彩色玻璃替代用彩色玻璃浆做的本体着色玻璃。这种产品玻璃本身没有颜色，但夹在两层玻璃之间的薄膜是彩色的。

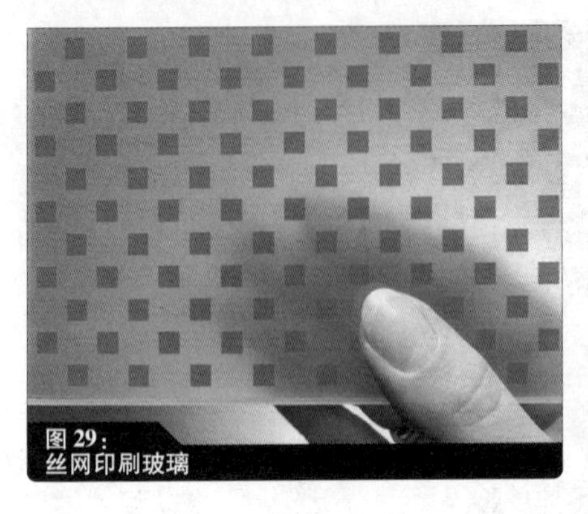
图 29：
丝网印刷玻璃

图 30：
使用丝网印刷玻璃的建筑立面

这种产品的构造是“玻璃—薄膜—玻璃”，由于中间彩色薄膜的原材料是聚乙烯醇缩丁醛树脂（PVB），因此其功能类似于传统的夹层安全玻璃。两片玻璃之间最多可以夹四层薄膜。因此就可以使用十一种基础色做出上千种透明、半透明或不透明的色彩。夹入夹层安全玻璃中间的薄膜不单有单色或有图案的，还有高分辨率数字印刷的。不过这种玻璃的色彩比丝网印刷的更为透明。〉见图 31

全息光学组件

全息光学组件（HOE）是指夹有全息格栅薄膜夹层的夹层玻璃。HOE 与棱镜玻璃的效果类似，会将白光分解为光谱色。颜色的效果取决于入射角度和观看角度。与双色玻璃类似，该产品可以产生动态的色彩效果。要将全息光学薄膜用于室外就必须将其夹入夹层玻璃保护起来。除了获得色彩效果，还可以用它来引入阳光以及进行阳光控制。另外，它还可以将阳光聚集到太阳能电池上，所以也用于提高光伏发电量。

图 31：
夹层有图案的夹层安全玻璃

提示：

施釉会使钢化玻璃的抗弯强度降低大约40%。某一表面最多可以施以四种不同的色彩。最大的施釉面积为3m ×6m。由于定制图形需要制作特别的样板，因此费用较高。施釉的那层表面耐刮擦且耐候，但会因长时间的紫外辐射而变色。因此，中空玻璃的施釉面一般为第二层表面。彩釉的色彩和数量会改变玻璃的太阳辐射得热系数，因此丝网印刷玻璃也被当作阳光控制玻璃。

提示：

彩色夹层安全玻璃一般比本体着色玻璃的吸热少；但制造某些深色产品时最好还是使用热处理玻璃原片。由于受到外层的保护，其色彩在紫外辐射下一般稳定不变。

P44

## 构造和组装

P44

### 概述

由于玻璃易碎，在玻璃结构的设计和施工中必须谨慎对待，详尽了解材料特性。与木、钢等很多的韧性材料不同，玻璃撞到硬物时会即刻破碎。因此，玻璃构件和玻璃结构的承压力需要经过严格的实验室检测。其中包括了对玻璃残余承载力——即玻璃发生破裂后其稳定性和承压力水平——的检测。有大量玻璃及玻璃构造方面的标准、技术导则和规范。在很多国家，如德国，每个建造非标准结构或为其生产构件的项目都需要获得针对项目的单独审批。

P44

### 框支承玻璃

需沿某一边沿连续固定的玻璃面板是框支承式的。在多数如窗、外墙面、玻璃屋顶等的情况下，面板玻璃是逐边支承的。也有沿三边、两边或单边支承的。〉见图32 底边连接屋面的全玻璃栏板就是一个单边支承系统的例子。〉见“应用”一章，“防坠人玻璃”一节

垂直玻璃面及玻璃采光顶

垂直玻璃面（与竖直方向夹角小于10°）和玻璃采光顶（与竖直方向夹角大于10°）是两个基本的大类。通常采光顶都要使用夹层安全玻璃（LSG）。PVB膜可以在玻璃破碎时起到防止碎片掉落的作用；用浮法玻璃或半钢化玻璃（HSG）制成的夹层安全玻璃（LSG）比单片玻璃的残余承载力大很多。如已采取某些防止大块的玻璃碎片掉落在公共通道的措施，譬如安装网格孔隙小于40mm的防落网时，也可使用单片玻璃（如浮法玻璃、钢化玻璃、压花玻璃）。

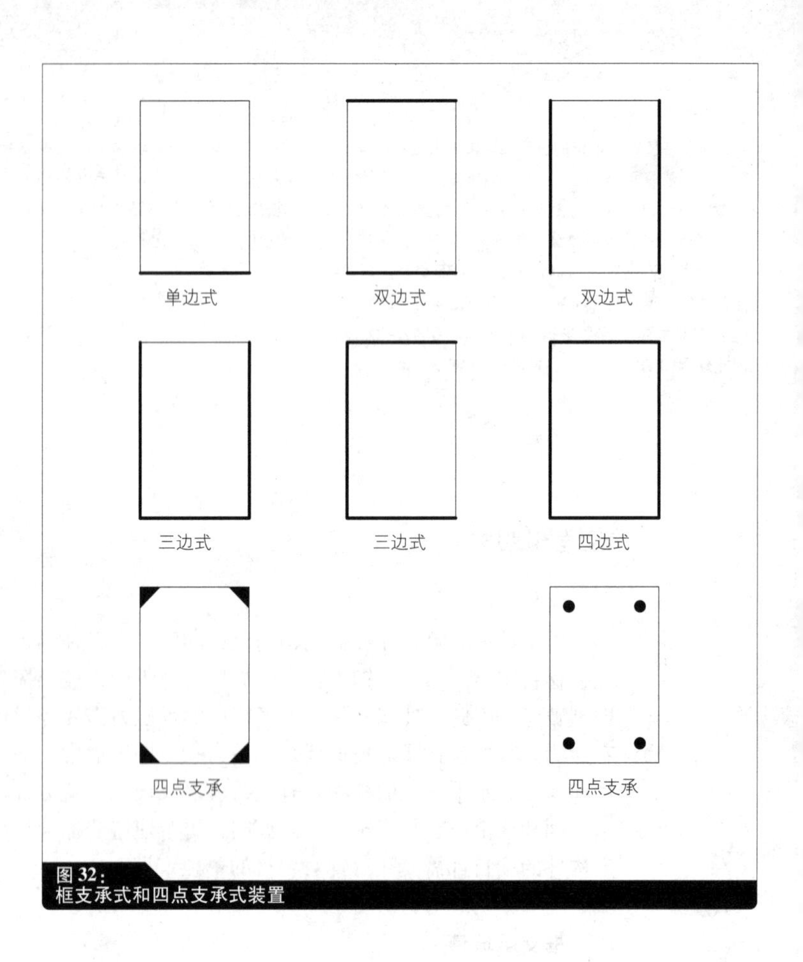


图 32：
框支承式和四点支承式装置

**提示：**

用 TSG 制成的 LSG 具有很高的弯曲强度，但在某些国家不许将这种玻璃用于上空区域，因为它的残余承载力很差。在上空区域使用中空玻璃时，底层玻璃采用 LSG。当主承载方向跨度不小于 0.7m 时，可以在上空区域使用夹丝玻璃。

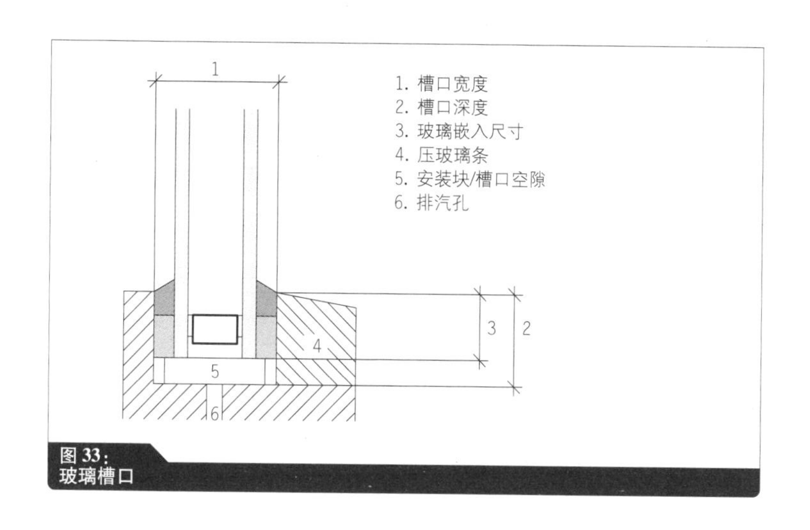


图 33：
玻璃槽口

P45

玻璃槽口

### 玻璃与框架

支座需要有柔性间层，以保证面板玻璃内受力均匀并补偿不均匀受力。无论如何，应避免玻璃片与钢、混凝土之类硬质建材的直接接触。玻璃嵌入尺寸特指玻璃伸入窗户槽口的深度。它由面板玻璃尺寸、允许偏差及结构的估算位移确定。〉见图 33

安装定位块的放置

在玻璃槽口中的支承销将玻璃自重荷载传到窗扇或框架，而框架中的安装定位块保证玻璃片不会侧移。〉见图 34

密封

现在的槽口做法一般不使用合成密封胶，这就保证能消除槽口内的汽压应力（减小应力）。嵌缝有两种方法：在密封条（硅胶、丙烯酸酯、聚硫或聚氨酯密封条）上面涂湿的密封胶；或者对使用压型条（如合成橡胶密封垫）的，则使用干的密封胶。必须保证槽口内积聚的冷凝水可以通过小小的排汽口蒸发散出去。〉见图 35

在设计窗或玻璃墙面结构时，应注意考虑不同合成密封材料的相容性。合成密封材料有基本的五个大类：丁基类、丙烯酸酯类、聚硫类、聚氨酯类及硅胶类。合成密封材料的化学组分各不相同，如，在稀释剂、溶剂、交叉胶粘剂及填料方面都不同。

压玻璃条

玻璃的框支承有两种基本形式。在窗户和立面元素的施工中，压玻璃条是最常使用的玻璃固定方法。压玻璃条安装于窗框的内侧。安装的方式或是钉在支承结构（如木框）上，或是夹在（如金属或塑料的框）上面。接触压力确保玻璃安装妥帖，并使密封紧实。

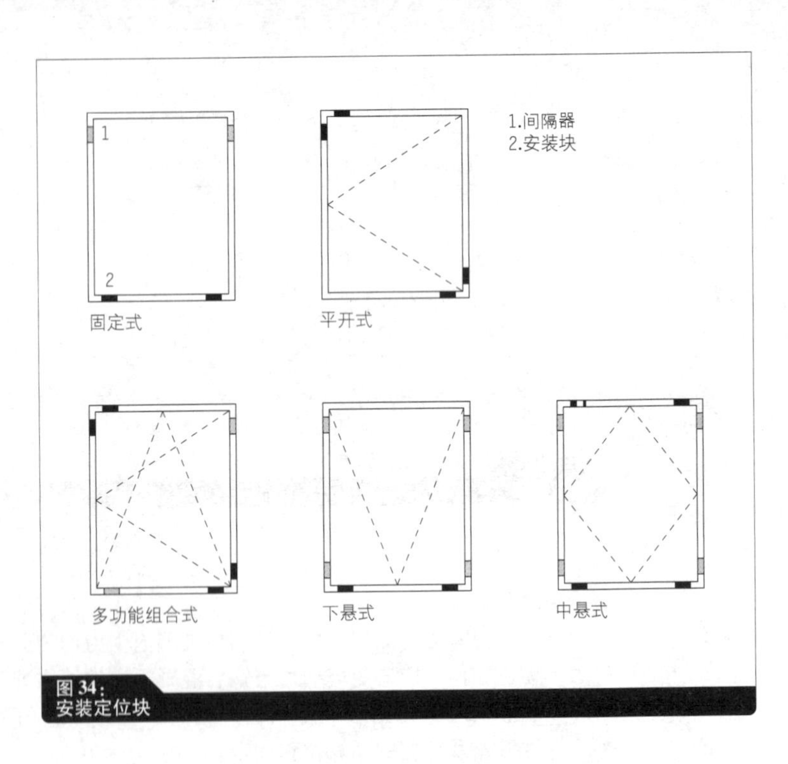

图 34：
安装定位块

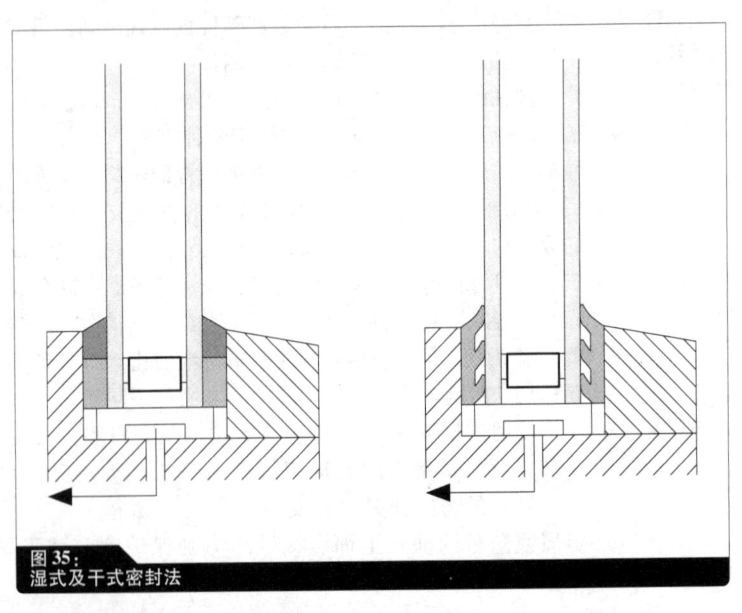
图 35：
湿式及干式密封法

**提示：**

错误的密封材料组合会损害安装定位块、边条或密封材料。为保证窗构造持久耐用，应核对厂商提供的合成密封材料相容性说明。

压框

压框是安装于外侧并将玻璃压紧于次结构的压条，材质有铝、钢、木及塑料等。这些密封条用螺丝安装，以使接触压力的位置准确。在多数情况下，螺丝上还要再盖上另一个压条。加上硅胶或EP-DM/APTK的耐候弹性密封条可以达到密封效果。如果窗户使用了中空玻璃，压框和支承框之间还要有绝热材料，如使用塑料隔热条等。在窗槽口内，玻璃之间的接缝也可能会积聚冷凝水，或者会有雨水从缝隙里钻进来。因此，需要一些小孔来消除蒸汽压。在规模较大的外墙施工中，水平与竖直接缝相贯通以形成连通排水系统。

压框可以指定使用与支承型材一样的材料，只是需要权衡不同材料的利与弊，例如，铝的抗腐蚀性比钢好，且使用冲压成型方式较易生产。不过，它需要用铝质盖板防止腐蚀。

有一种特殊的压框叫做复合型压框。这种构件采用耐久塑性材料一体成型，兼具了压框和密封条的功能。〉见图36～图39

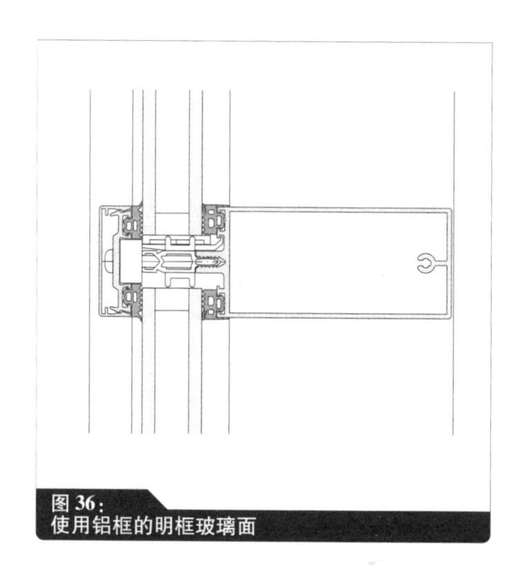

图36：
使用铝框的明框玻璃面

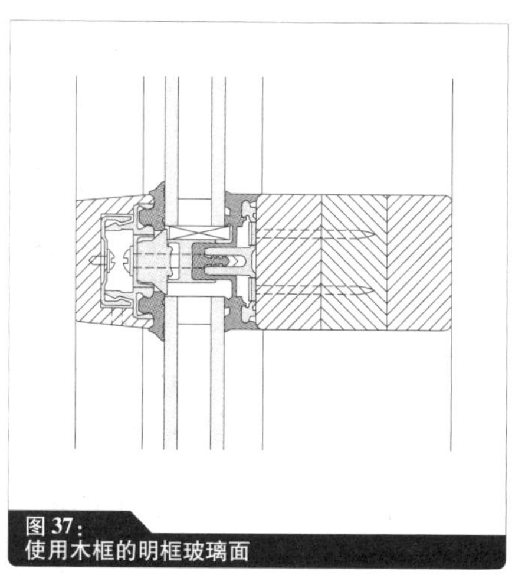

图37：
使用木框的明框玻璃面

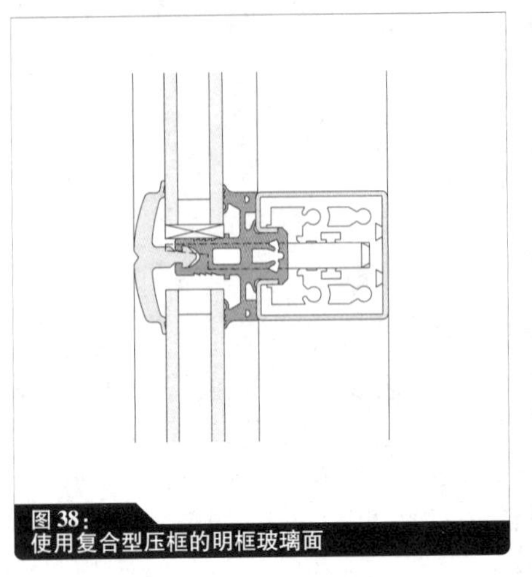
图 38：
使用复合型压框的明框玻璃面

图 39：
组合明框玻璃墙面

P49

**点支承玻璃**

这种安装方式的一个优点是能够产生精美、通透的玻璃面。由于采用点式支承固定，面板玻璃只在少数几个点受力而不是整条边都受力。矩形或方形的面板玻璃至少要有四个角固定；大片的玻璃则以点固定支承于次结构上。由于局部应力可能非常大，一般建议使用热处理玻璃（钢化玻璃或半钢化玻璃），有些情况下根据规定必须使用热处理玻璃。有两种支承紧固件：不需要穿透玻璃的无孔式的，以及要在玻璃上穿洞的。

无孔式支承

无孔式支承靠夹住玻璃的角和（或）边来固定住玻璃片。材料采用铝或不锈钢，有从方形到圆形的若干不同形状。〉见图 40 ~ 图 44

在设计及施工过程中，必须注意要在金属与玻璃之间使用弹性垫层，以避免二者有任何的直接接触。倾斜或由支架产生的太大的接触压力都会造成玻璃破碎，因此这些情况应尽可能避免。无孔式墙面的夹具往往是为项目个别定制的。根据夹具不同，玻璃片或平齐安装或鳞式安装。玻璃表面的夹片覆盖面积不小于 $1000mm^2$（且槽口深不小于 25mm），且该面积的大小由玻璃的计算受力决定。由于这种方式不用在玻璃片上钻孔，夹片要在玻璃的接缝位置用螺栓固定。

点式玻璃的夹片也可与框式支承相结合。使用这种方法时，面板玻璃置于连续的支承框上，每隔一定间距用一个夹片将玻璃固定。

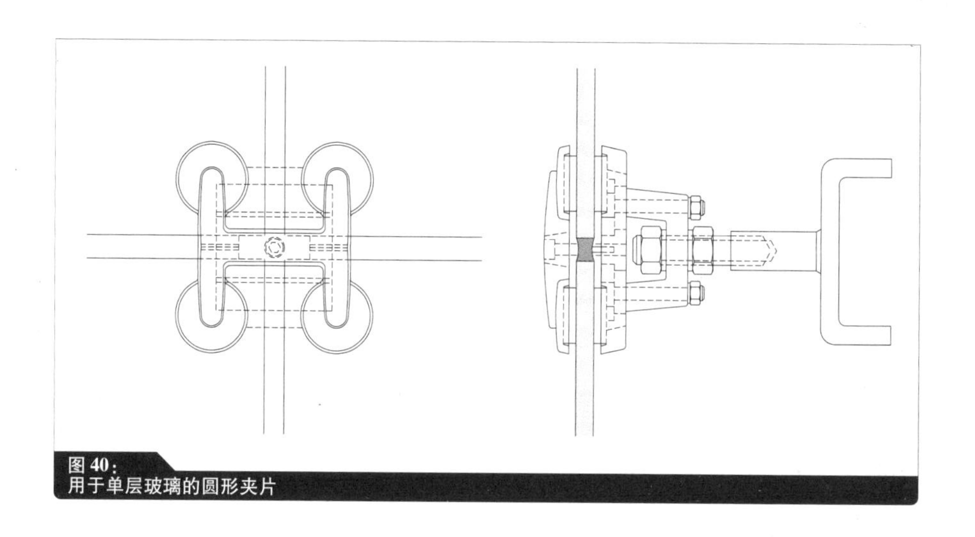

**图 40：**
**用于单层玻璃的圆形夹片**

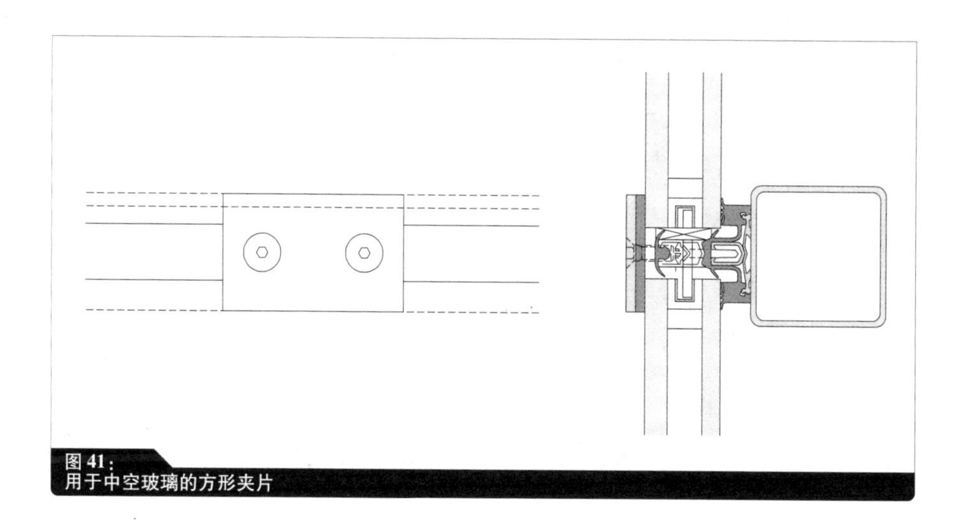

**图 41：**
**用于中空玻璃的方形夹片**

有孔点支承

有孔点支承的方式在玻璃平面内固定玻璃。由于需要开孔，这种方法比无孔式支承需要更多的人力投入。无孔点式安装有各种形式：平齐安装（平头式）及附有压片（夹片）的点式安装，如球面形（半沉头）或圆柱形。〉见图 45 ~ 图 48

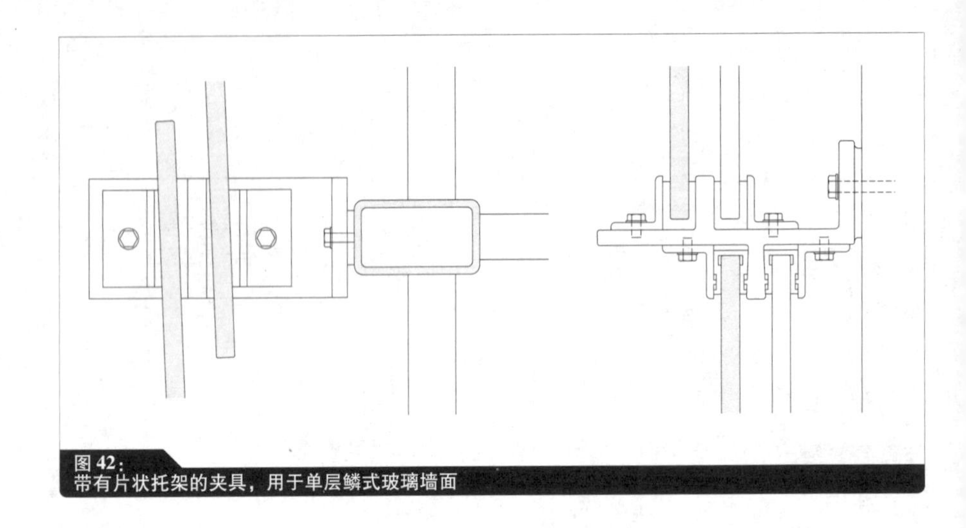

图 42：
带有片状托架的夹具，用于单层鳞式玻璃墙面

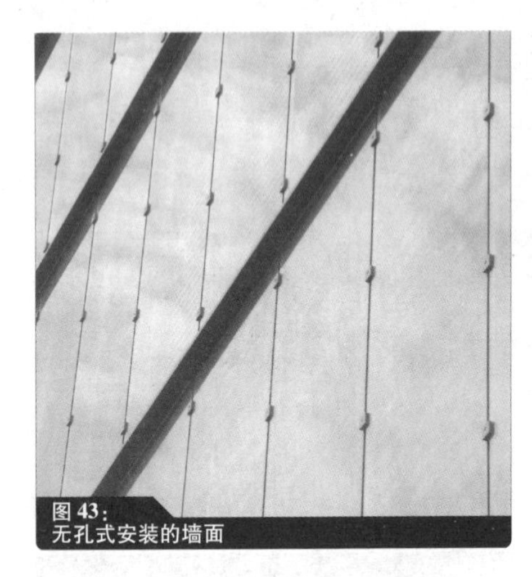

图 43：
无孔式安装的墙面

图 44：
无孔式安装的鳞式墙面

由于在玻璃孔附近的表面已经因开孔而破损削弱，同时也因为这一区域必须承受最大的应力，因而这种固定方式要求系统为非常稳定的静力状态。孔与孔之间、孔与玻璃边沿之间的最小距离不小于80mm。对中空玻璃，在钻孔周围还需要另外密封以确保玻璃原片之间不漏水。有孔式点支承方法的研发费时费力；一般这种玻璃墙面或采光顶会采用已获专利的安全模式支承固定。

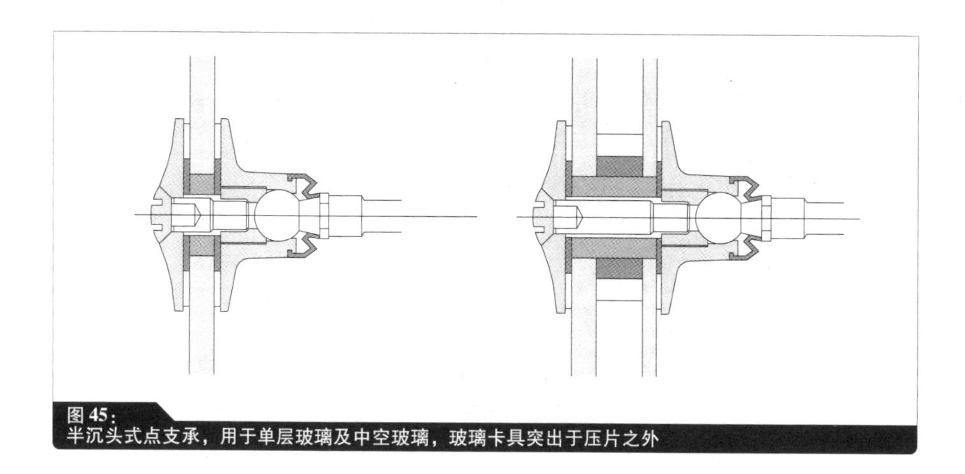

图 45：
半沉头式点支承，用于单层玻璃及中空玻璃，玻璃卡具突出于压片之外

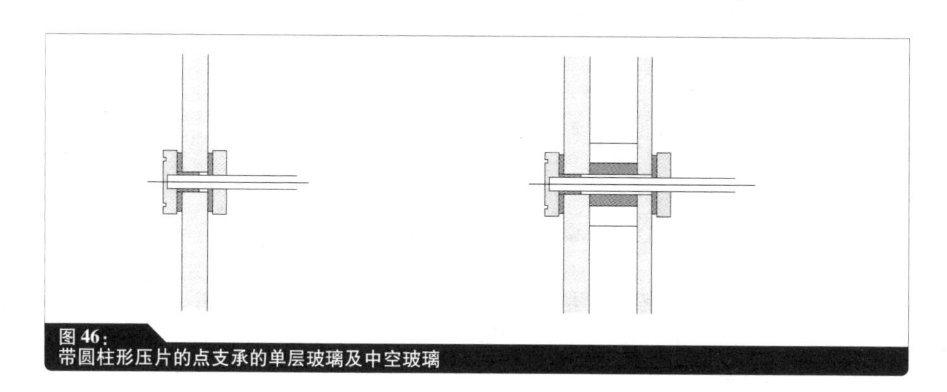

图 46：
带圆柱形压片的点支承的单层玻璃及中空玻璃

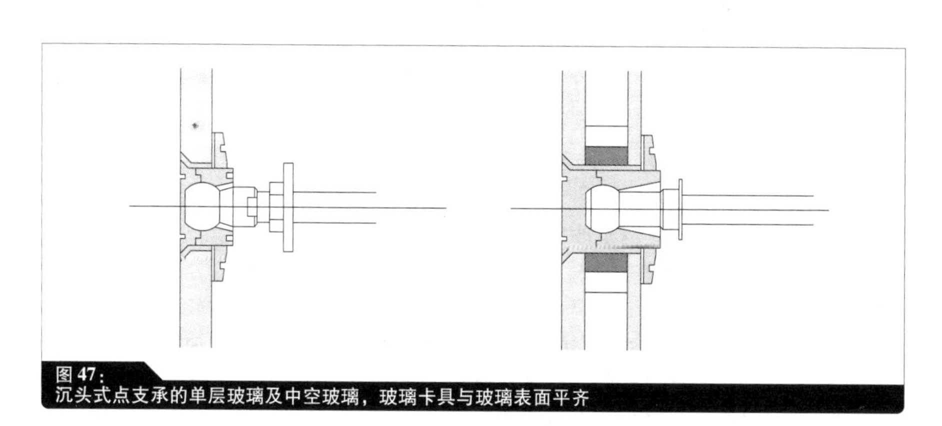

图 47：
沉头式点支承的单层玻璃及中空玻璃，玻璃卡具与玻璃表面平齐

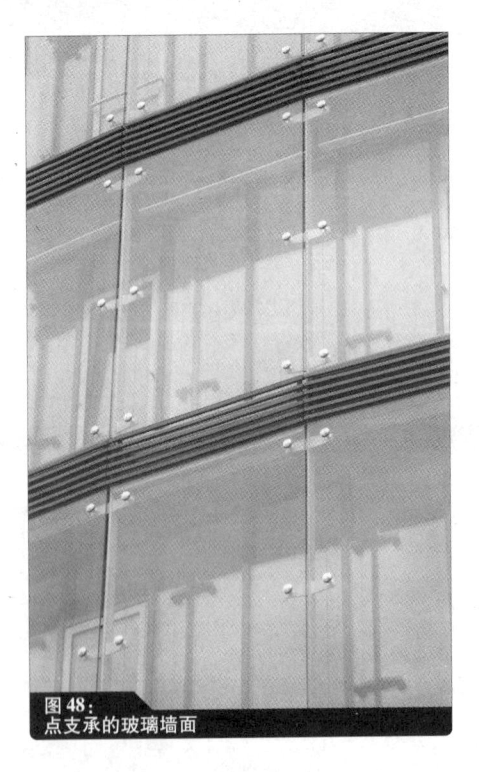

图 48：
点支承的玻璃墙面

沉割锚固

“沉割锚固”是点式安装的一种特殊形式。这种方式只从玻璃的一侧固定，而不需要压紧另一侧。将柔性的圆柱形配件安在玻璃上的圆锥形孔内，通过夹住这个圆柱来支承固定玻璃。〉见图 49

玻璃墙面常使用复合球铰式驳接头，这种装置有助于避免固定点附近较大的玻璃弯曲应力。点支构件由柱头螺栓固定于次结构上。次结构和柱头螺栓应便于接下来的玻璃安装并能补偿尺寸偏差。像带缝槽的平头型钢，或者特殊的不锈钢四点支承装置之类都较适合这种情况。〉见图 50

P54

玻璃接缝

**玻璃接缝及玻璃转角**

如果没有使用盖缝构件，点式支承的玻璃边缘会外露。对气候要求不高的单层玻璃系统，如用于多层车库或仓库的，玻璃接缝可以这样外露。这种方式的施工较为经济且有利通风。但对中空玻璃墙面则必须密封其接缝。且考虑到玻璃的热胀（扩张）性能以及立面的受风弯曲，接缝应为弹性的。

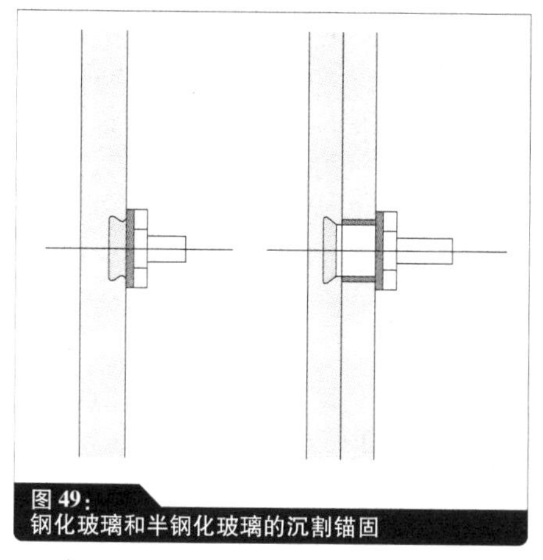

图 49：
钢化玻璃和半钢化玻璃的沉割锚固

图 50：
四点驳接件

密封条

主要有两大类密封条。第一种是将 EPDM/APTK 或硅胶类的密封条塞在接缝处。第二种是从外侧将密封胶喷涂到密封条上，从外面将接缝密封住。〉见图 51，图 52 对中空玻璃和夹层玻璃，槽口应保持开放，这样便于消除蒸汽压和积水，保证玻璃边缘的耐久性。这种方式使得从缝隙处漏进的水和冷凝水可以经由排水系统很快排出，而不至于太长时间地积于一处。外露的中空玻璃封边必须有耐紫外的性能。可以用硅胶（不要用聚硫橡胶、聚苯乙烯）作为第二道密封，也可以将（一般为黑色的）涂塑条塞到外层玻璃的内表面（即第二层表面）并盖住接缝密封剂，这样都可以形成耐紫外的封边。

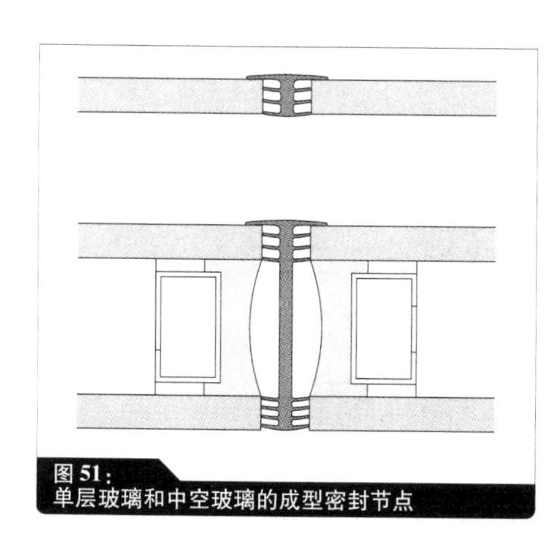

图 51：
单层玻璃和中空玻璃的成型密封节点

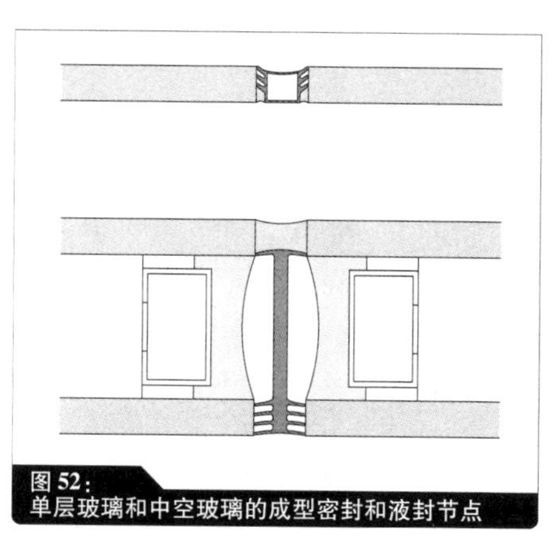

图 52：
单层玻璃和中空玻璃的成型密封和液封节点

玻璃转角

处理玻璃转角的原理与玻璃接缝相近似。必须注意及确保消除蒸汽压以及密封材料的化学性能相容。与玻璃自身相比，转角处的隔热要求相对较低，因而需要考虑冷凝水聚积的问题。对中空玻璃无框式立面，有多种可行的转角构造方法。以下几种最为常见：

— 不透明转角：转角处填以隔热条，〉见图 53

— 斜角错接的中空玻璃，〉见图 54

— 对缝错接的中空玻璃。〉见图 55

错接式，特别是斜角错接式的隔热转角施工较为复杂，组件和节点都要精准。不过其优点在于包括转角在内都采用同一材料，视觉效果一气呵成。

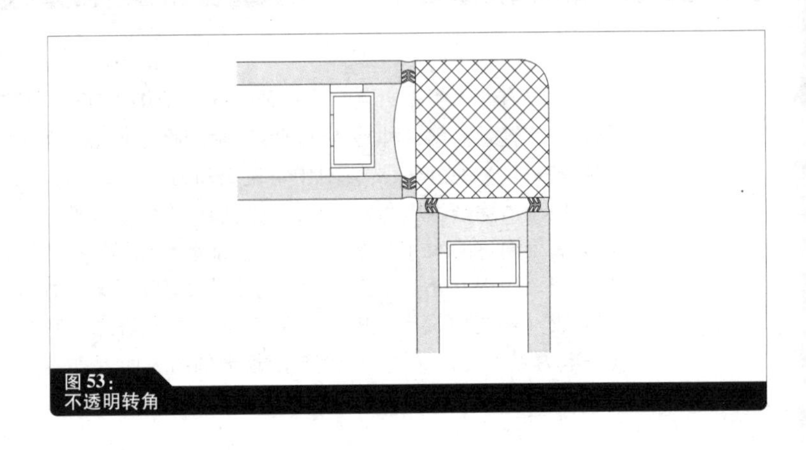
图 53：
不透明转角

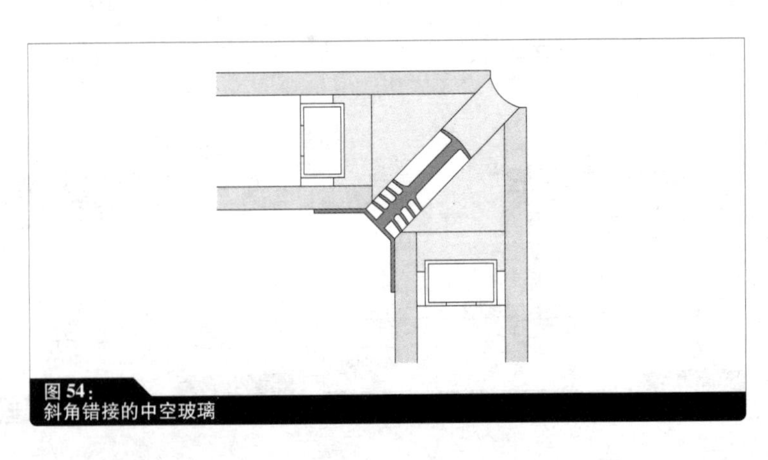
图 54：
斜角错接的中空玻璃

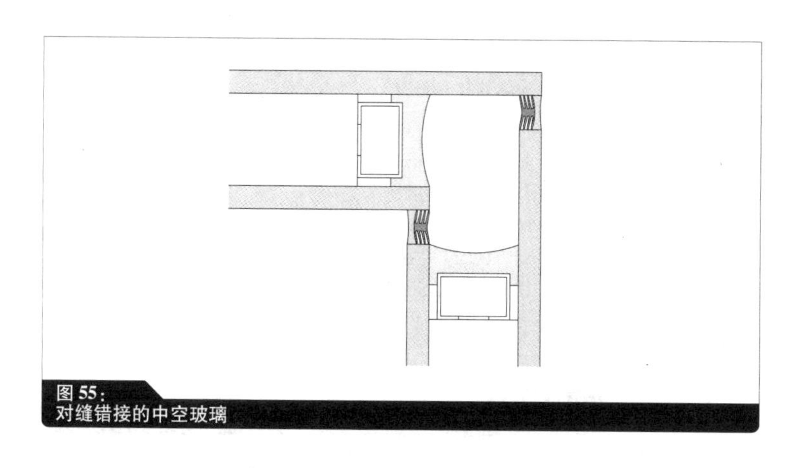

图 55：
对缝错接的中空玻璃

P55

**结构密封胶粘式玻璃**（SSG）

结构密封胶粘式构造是以胶粘剂粘合结玻璃而成。它是框支承方式的特殊形式，需要申请批准才能采用，因此需要在建筑法规指导下取得针对所选构造形式的单独批文。由于这种构造是将玻璃片粘于金属框架（转接框）上，金属框架再固定于支承框，因而形成了平整的无框表面。

注胶

注胶可抵御风荷载，玻璃的重量由“常规”的安装定位块系统承担。全方位注胶不能在建造现场实施。它需要一个具备资格的厂家在精确的温湿度及洁净无尘的条件下进行。通常是由玻璃厂在生产玻璃之后直接实施注胶。玻璃片必须非常清洁、干燥且无油。

抗负风压措施

在某些国家，如德国，高度超过 8m 的玻璃墙面需要有专门抗负风压的金属的安全装置。这样可以在粘结力失效时防止玻璃坠落。对注胶的要求是很严格的，必须能承受各种荷载，包括温度、湿度、紫外线照射及微生物腐蚀引起的各种变化。需要特别的抗负风压装置是因为很难确保注胶粘结构造的长期承载力。抗负压装置可以采取类似全面框架的支承或点式支承。SSG 立面在其注胶、抗负风压装置、接缝密封及中空玻璃的具体构造上有若干不同的系统形式，可归为错缝式（错缝中空玻璃）与平缝式两类。〉见图 56 · 图 58

胶粘剂要能与硅胶、聚氨酯及其他化学物质相容。与点式支承的玻璃墙面一样，SSG 玻璃墙面上外露的胶粘剂也需要防紫外线。在 SSG 中使用的粘结构造与全玻璃结构一样日益普及。这种构造还能承担重要的结构功能。〉见“应用”一章

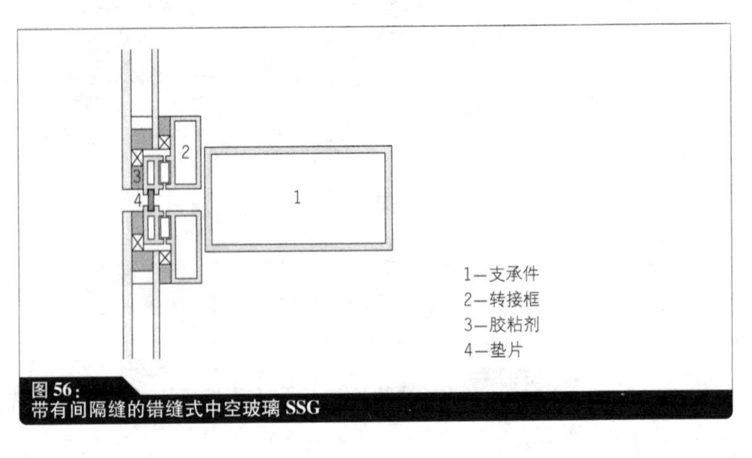

图 56：
带有间隔缝的错缝式中空玻璃 SSG

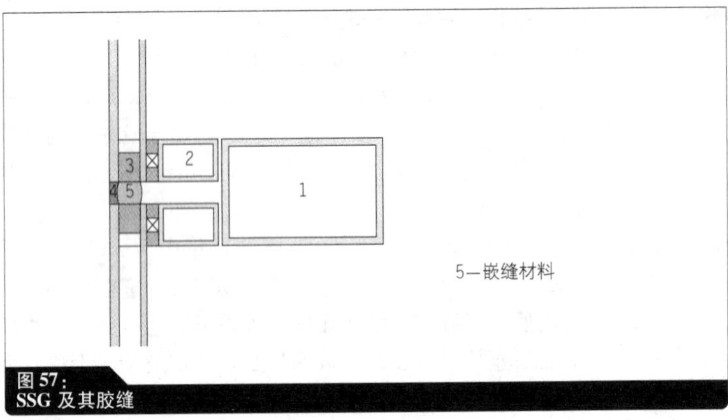

图 57：
SSG 及其胶缝

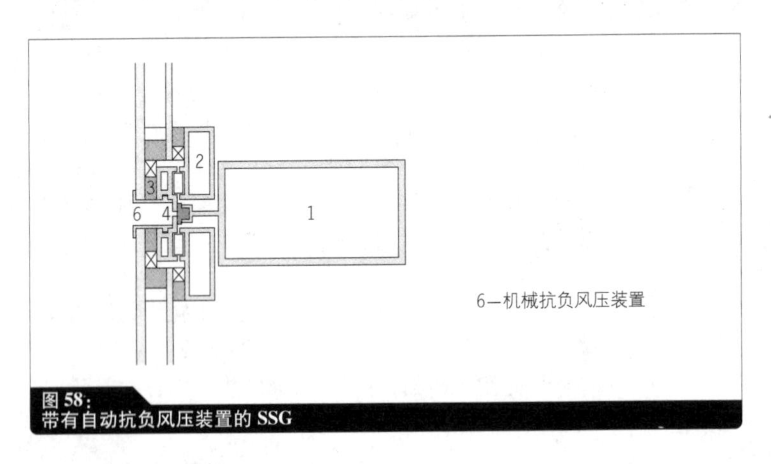

图 58：
带有自动抗负风压装置的 SSG

P59

# 应用

P59

**垂直玻璃面**

尽管只有特定种类的玻璃可以用于玻璃采光顶〉见“构造和组件”一章，但理论上，任何种类的玻璃都可以用于垂直玻璃面。不过实际上，针对玻璃的安装位置、用途及构造，仍有一些限制。

玻璃类型，破坏的风险

钢化玻璃（TSG）或夹层安全玻璃（LSG）经常替代浮法玻璃使用，以尽量降低发生事故的风险。例如在学校和幼儿园的人流区域，如未采取栏杆或扶手之类的其他措施，则需要采用安全玻璃以防有人撞碎玻璃。有时中空玻璃的外层玻璃也采用钢化玻璃以减小玻璃破碎的风险：特别是在公共人流区域上方的立面或不能采用逐边固定玻璃时。对单层玻璃而言，若非逐边固定，最好使用钢化、半钢化玻璃或夹层安全玻璃，或者如有必要使用夹丝玻璃，而不用浮法玻璃。很多国家都有该项规定。当玻璃安装于4m高度以上时，不使用普通的钢化安全玻璃，而是要用经均质处理（均热）的钢化玻璃，以避免硫化镍杂质造成的“自爆”。〉见“建材玻璃”一章

〉✎

点支承玻璃一般要求使用热处理玻璃。当安装于4m高度以上时，一般使用两片钢化玻璃或半钢化玻璃制成的夹层安全玻璃，以此防止玻璃板块因损坏而坠落。

承载力，组件尺寸

除了要考虑设计及功能的要求，组件尺寸还取决于荷载、固定方式的类型以及所用玻璃的类型。计算垂直玻璃的构件时，除去构件本身的重量，风荷载（正负风压）、气候（对中空玻璃而言）〉见“建材玻璃”一章、特别是交通荷载（如，对商店窗户的水平冲击）也必须考虑。理论上，使用逐边支承的方法可以安装最大3.21m×6.00m（带型）的单层、甚至是中空玻璃。对于固定边少于4边的框支承或对于点支承，最大规格则会小很多，因为玻璃内部必须承受更大的弯曲及力。玻璃的重量更会限制尺寸：一块3m×6m的中空玻璃常会重

✎

**注释：**

有高吸收率（太阳控制）膜的玻璃或本体着色玻璃在阳光下比普通无色玻璃升温更多，因此应采用热处理玻璃。首先应该注意那些热辐射吸收率在50%以上的玻璃。

图 59：
施工现场正在安装大片的中空玻璃

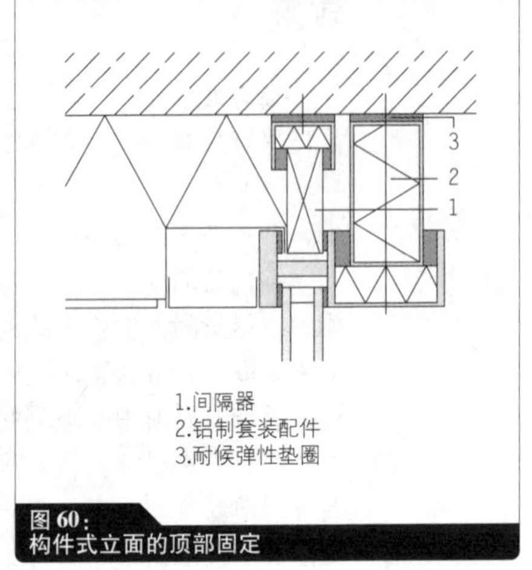


图 60：
构件式立面的顶部固定

达 2 吨。安装又大又重的玻璃面板很复杂，且需用重型真空起重机举起玻璃。〉见图 59

墙面类型　　玻璃墙面大体可以划分为两类：构件式立面和单元式立面。

所谓“构件式”是指墙面的支承结构由竖向构件（柱）和横向构件（梁）组成。一根根横梁和立柱在工地焊起来或用螺丝拧紧，组装起来。然后从外侧将玻璃固定到该结构上。采用构件式结构可做成更大的跨度，但它的缺点在于现场组装复杂，比“单元式”更耗时。

单元式墙面的所有结构构件，如框架、玻璃、窗扇都是在工厂中组装成预制单元的，墙面就由这些预制单元构成。〉见“应用”一章，“开窗”一节预制单元的尺寸必须适宜运输，这就限制了规格方面的选择。

支承的细节　　玻璃与相邻建筑局部如屋顶或地面等的连接，不允许给相应的玻璃单元造成额外的荷载。

顶部固定　　为此，要用一个塑性间隔件或铝制套装配件将玻璃组件单元连接于顶棚。交通荷载或安装造成的屋顶位移可以由这个柔性固定来消除。这样，就保护了墙面的密封和隔热构件。〉见图 60

底部固定　　间隔件也用于底部固定的构件式立面；它与防水材料一起连到立面底部的梁上，固定于压框的后面。需要特别注意，密封材料和玻璃立面连接的位置应至少高于浸水层 15cm。如果外墙使用上部固定或底部固定的形式，就必须有水槽以保证排水的顺畅。〉见图 61

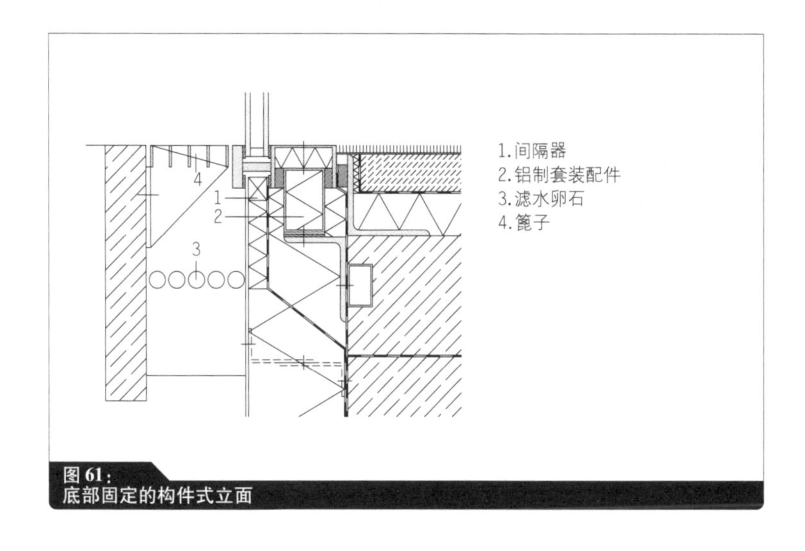


图 61：
底部固定的构件式立面

P61

**开窗**

建筑外表面的开窗有一些基本的功能要求。此外，其设计对整体的形象非常重要。对垂直玻璃面而言，开窗的重要性还在于调节建筑内部微气候。开窗使室内空气自然流通，有助于防止空气因太阳辐射而持续加温。除了传统的平开窗，玻璃墙面及玻璃屋顶还使用其他形式的开窗来保证持续通风；它们能根据室温自动控制。下面列出并描述了各种玻璃面上的开窗：〉见图 62

墙上的窗构件是最简单基本的立面原形。它可以是单块构成，也可以是分为几块的，这样就可以既有固定窗又有可开启窗（开启扇）。

天窗向屋顶下面提供光线和风。所用玻璃必须符合对采光顶的要求。天窗一定要有倾斜角度来保证无组织排水的顺畅。〉见图 63

玻璃百叶或百叶窗实现了对通风截面的准确控制，因此调节了室内的通风。百叶形式有多种，有框式和无框式，对应于单层玻璃和中空玻璃。〉见图 64，图 65

以上讨论的开窗形式也可与明框玻璃墙面组合应用。开窗组件的窗框可以用支承框架和压框替代，在外观上会很明显。

结构密封胶粘式玻璃墙面（SSG）上使用的是外开上悬窗。这种做法使窗看上去就像无框的，效果很漂亮。〉见图 66

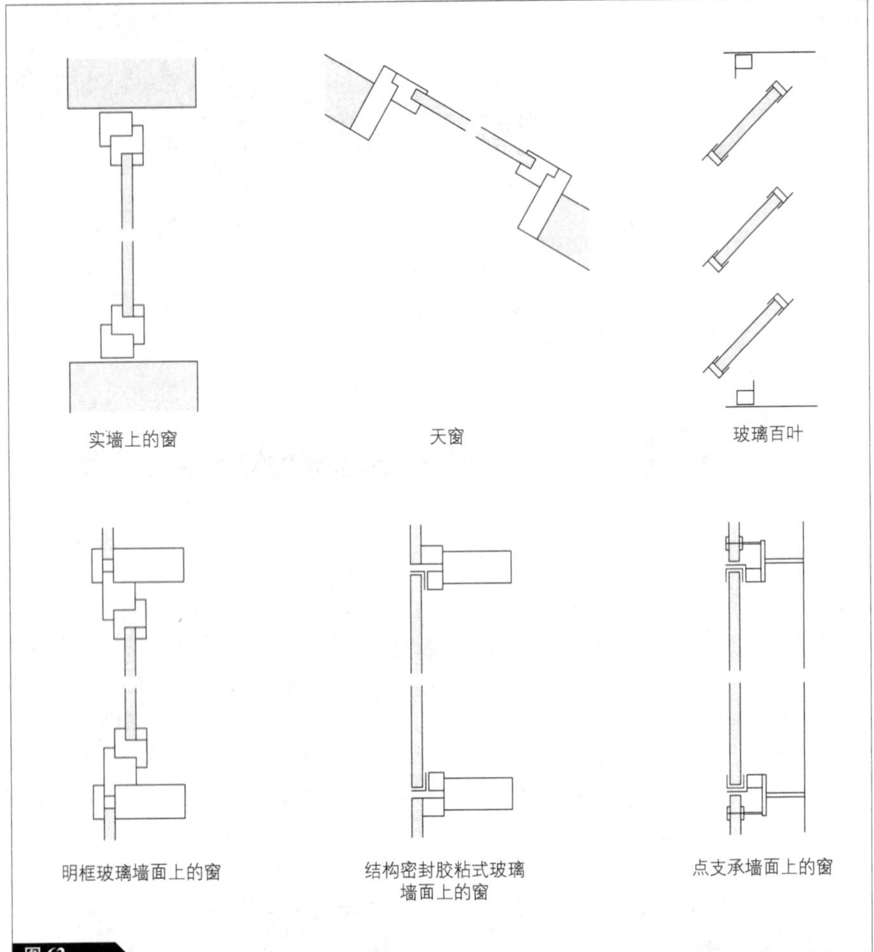


**图 62：**
**开窗组件的一般分类**

**注释：**

关于开窗方面的深入讲解可以参考本套丛书中的《窗户设计与施工》一书，罗兰德·克里普纳与弗罗里安·穆索编著，中国建筑工业出版社预计于 2010 年 3 月出版。

**注释：**

使用单层玻璃百叶是为了获得可调式的外遮阳。为此，玻璃表面有印刷或镀膜。如果加上太阳能电池组件，则可发电来调节玻璃百叶。

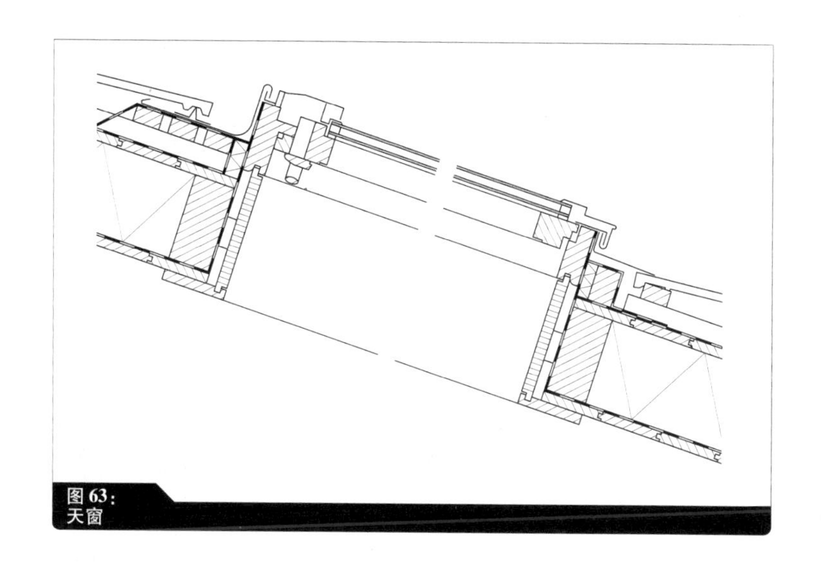

图 63：
天窗

在点支承玻璃墙面上加入窗组件是个特殊的挑战，因为要保持立面的精美效果，即便在开窗处也不例外。比如，对单层玻璃墙面，可使用上悬窗扇或点支承的玻璃百叶。

中空玻璃墙面需要更加精心的设计。因为没有框架，保证隔热效果的那些密封和固定配件很难连接到墙面上，可以选择像 SSG 表面一样安装上悬窗扇。

P65

**防坠人玻璃**

取代扶手或栏板使用、并具有建筑法规规定的防止人坠落的功能的玻璃结构称为“防坠人玻璃”。其应用形式多样，从玻璃栏杆嵌板，到全玻璃栏杆，再到房间通高的框支承或点支承玻璃。在很多国家，防坠人玻璃的安全性需经“摆式撞击测试”之类的动态压力实验来证实。〉见图 67

由于这种实验非常复杂，结构设计时应采用经检验证明可靠的构造经验，以避免受到进一步的压力检测。防坠人玻璃可分为三大类：

固定式通高玻璃面

第一类为固定式通高玻璃面——即玻璃面上没有承受水平力的开启扇、扶手或突出的横杆等。从结构的角度看，逐边支承是最有效的解决方案。如不适用该方案，则必须以其他方式保护非固定边免受撞击，如利用相邻的玻璃板或邻近的墙、顶棚等建筑构件。两边固定的玻璃有两个自由边，因而受撞击时弯曲变形会大得多。这就要注意使

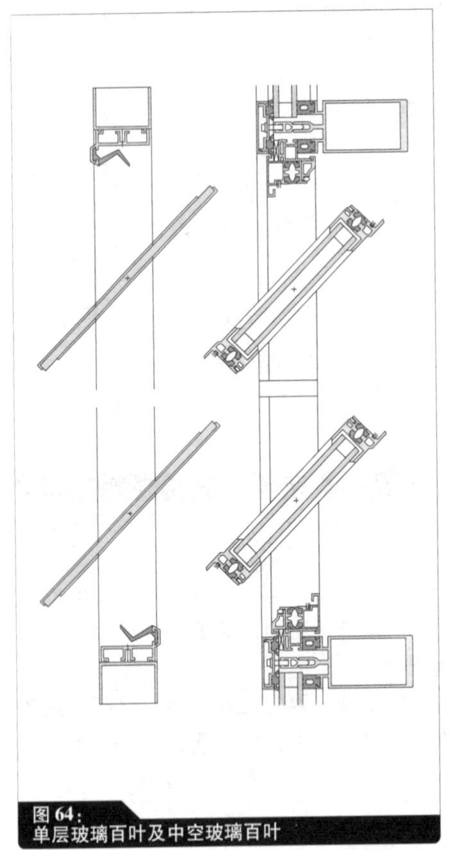

图 64：
单层玻璃百叶及中空玻璃百叶

图 65：
使用玻璃百叶的窗

玻璃嵌入足够的深度。这样玻璃就不会滑出支座。点支承玻璃的夹片直径不应小于 50mm。

**提示：**

在多数情况下，防坠人玻璃需要使用夹层安全玻璃，对点支承式要使用钢化、半钢化玻璃做的夹层安全玻璃。当面板破碎时，玻璃夹胶和高抗拉性薄膜仍可以提供防人撞击的足够保护力。

**注释：**

一种特殊的情况是带有防撞扶手（如安装于内侧的圆形不锈钢配件）的通高玻璃，其扶手高度在建筑法规中有规定，撞击时扶手抵挡了绝大部分压力。这种情况下的玻璃厚度可以减小而外观不变。

图 66：
结构密封胶粘式玻璃墙面（SSG）上的窗

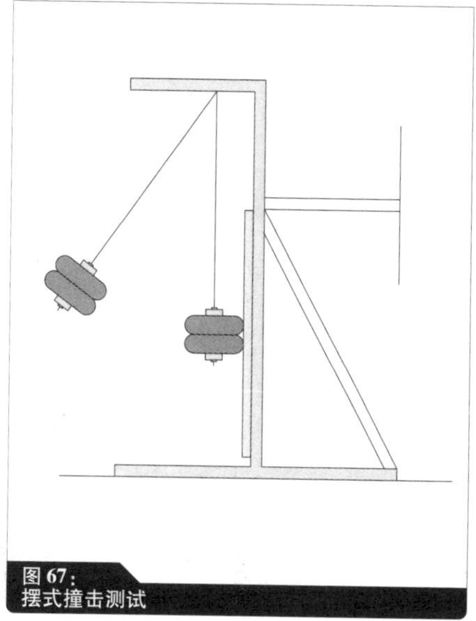
图 67：
摆式撞击测试

固定式受力玻璃栏板

第二类包括了在底部以夹式框支承构造固定的受力玻璃栏板。顶边需要防止受到撞击力，比如采用粘结式构件或连续固定的扶手等。后者应有适宜的尺寸，以保证单片玻璃失效时水平荷载会传递到下一片玻璃。这类方式以采用钢化、半钢化玻璃做成的夹层安全玻璃为宜。〉见图 68

玻璃嵌板

第三类就是将玻璃作为填充部分的立面扶手或玻璃栏板。水平荷载由受力栏杆或立面横杆承担。玻璃至少是对边点式支承的或框式支承的。面板玻璃是夹层安全玻璃或钢化玻璃。〉见图 69

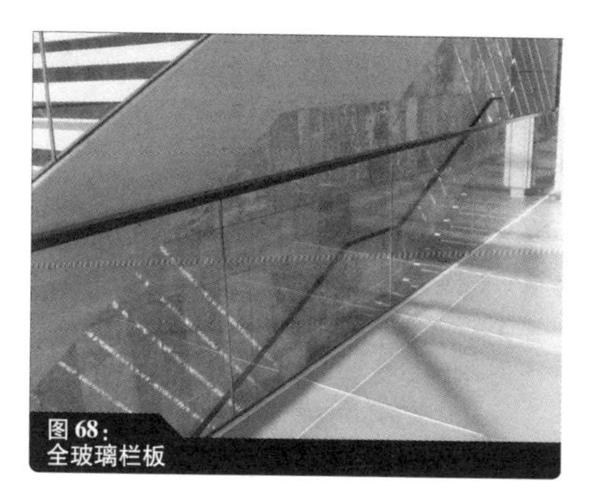
图 68：
全玻璃栏板

图 69：
有玻璃嵌板的栏杆

P67

**玻璃采光顶**

和垂直玻璃墙面一样，采光顶有框式支承和点式支承两种形式。

荷载，尺寸

采光顶上的结构荷载比垂直玻璃大，因为玻璃的自重垂直作用于板面上，而且风、气候及雪荷载等也需要考虑。由于受力较大，采光顶的组件规格不能像竖直面上的那么大。对较大的玻璃屋顶，还要考虑上面人和物的荷载，因为必须要做成上人采光顶，以保证维护和清洁工作的进行。〉见“应用”一章“限制型及开放型上人采光顶”一节

另外，玻璃采光顶必须抗撞击，比如要能承受落下的冰雹或树枝之类的东西。因此，采光顶的上层应该用热处理玻璃。对点支承采光顶而言，需要的厚度不仅取决于荷载，还取决于残余承载力的要求。对夹片也有这方面的要求。〉见表 7

屋顶排水

玻璃面需要向排水方向（即向天沟或邻近的平屋顶）倾斜足够的角度。压框（对明框式玻璃）的安装应有倾斜角度以利排水。〉见图 70

**表 7：**
**具有检定残余承载力的点支承采光顶，支承点垂直网格式排列**

| 夹片直径（mm） | 玻璃最小厚度（mm）HSG 制成的 LSG | 某一方向的固定点间距（mm） | 另一方向的固定点间距（mm） |
|---|---|---|---|
| 70 | 2×6 | 90 | 75 |
| 60 | 2×8 | 95 | 75 |
| 70 | 2×8 | 110 | 75 |
| 60 | 2×10 | 100 | 90 |
| 70 | 2×10 | 140 | 100 |

**提示：**

围合电梯或自动扶梯的玻璃还必须采取保证传送系统安全运行的措施。如采用特殊的安装方式使人无法越过或被夹在两片玻璃的缝隙处（无框式玻璃的情况下）。

**提示：**

点支承玻璃的最低要求为 2mm ×6mm 厚钢化玻璃夹最薄 1.52mm 厚 PVB 膜组成的夹层安全玻璃。夹片的最小直径为 60mm。自由端尺寸（即玻璃边缘与固定点之间的距离）最小 80mm，最大 300mm。

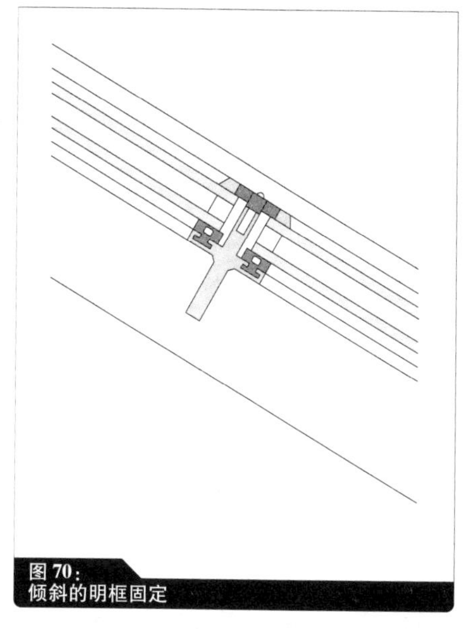
图 70：
倾斜的明框固定

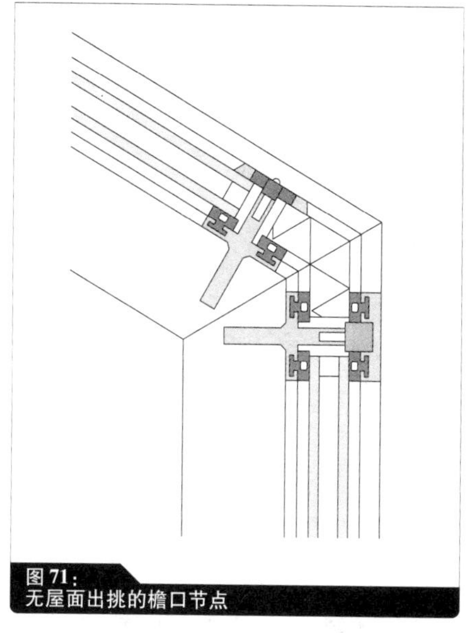
图 71：
无屋面出挑的檐口节点

与垂直玻璃面相连的檐口节点构造可以有出挑或无出挑。如果没有出挑，雨水会直接流向立面，造成污染。因此，较大的玻璃屋顶应该通过天沟排水。〉图 71 ~ 图 73

P69

限制型上人采光顶

**限制型及开放型上人采光顶**

为维修及清洁表面而必须上人的采光顶称为上人采光顶。尽管称为“上人”采光顶，但应注意玻璃面上只能同时承载有限的几个维修人员。除了要保证玻璃有一定的结构强度，还应设计一些防滑措施。屋顶倾斜角度超过 20°时，应安装安全钩。玻璃采用夹层安全玻璃，其抗撞击强度和残余承载力应验证合格。对中空玻璃而言，上层可以用钢化安全玻璃代替夹层安全玻璃。

开放型上人采光顶

开放型上人采光顶与限制型的不同，它还向公众开放，因此所受交通荷载要大得多。通常设计的荷载上限为 50KN/$m^2$。上人采光顶一般使用厚度不小于 30mm 的夹层安全玻璃。多由三层玻璃组成，最上面为防撞击的钢化或半钢化玻璃面，可以保护下面的两层。下面两层玻璃承重，因此即使上层玻璃损坏了，玻璃面仍有承受人站在上面的足够强度。〉见图 74

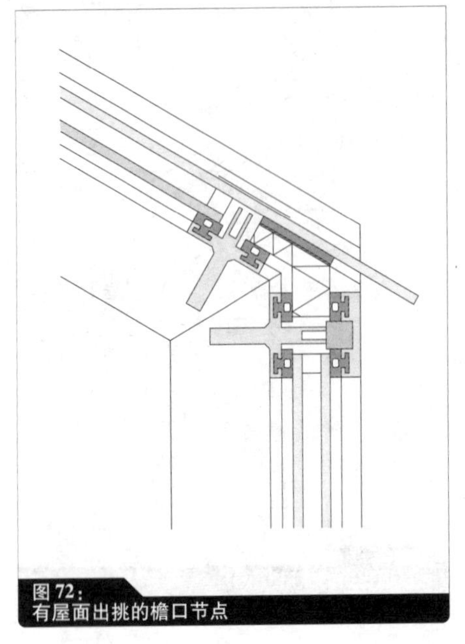

图 72：
有屋面出挑的檐口节点

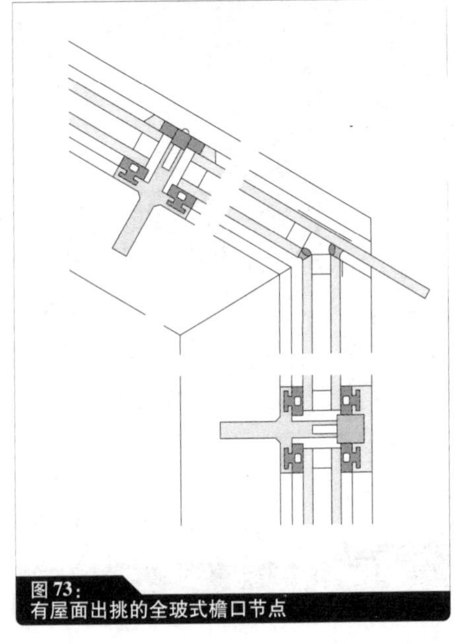

图 73：
有屋面出挑的全玻式檐口节点

防滑

这种玻璃平面表面应防滑，比如采用彩釉丝网印刷，使整个或局部的玻璃表面是粗糙的；也可以使用蚀刻达到表面粗糙防滑的效果。上人采光顶通常采用两边或四边固定的框支承方式。过去还必须使用螺栓固定于支承结构上。玻璃片嵌于抗压弹性体支撑上，嵌入的深度至少为 30mm。玻璃与玻璃、玻璃与金属之间隔以间隔件。可上人的采光顶适用于类似室内楼层的楼梯等地方。采用中空玻璃的上人采光顶构造较为复杂，因为与其他轻荷载上人采光顶不同，它不可以将太大的交通负荷传递到边缘构件上，因为它的边构件不应持续受力发生破坏。因此用于博物馆展示区域之类地方的轻荷载上人采光顶常采用双层屋面的构造。这样，构造上层可以是轻荷载上人采光顶，下层是满足隔热需要的中空玻璃。〉见图 75

P70

### U 形玻璃

U 形玻璃的优点是立面无需使用次级结构来承重。U 形玻璃刚性大，无需边框就能组成大面积的玻璃面。U 形玻璃可以竖向或横向组装。为了安装及传递荷载，要将 U 形玻璃嵌入铝框中约 50mm 深的窄缝内，其构造可以为单层或双层的形式。〉见图 76

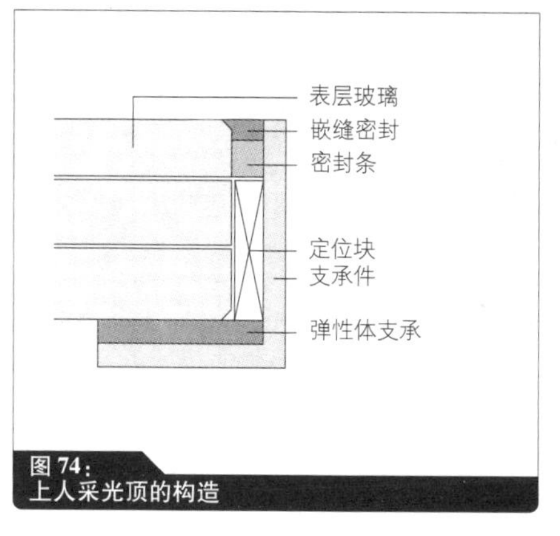


图 74：
上人采光顶的构造

图 75：
开放型上人采光顶

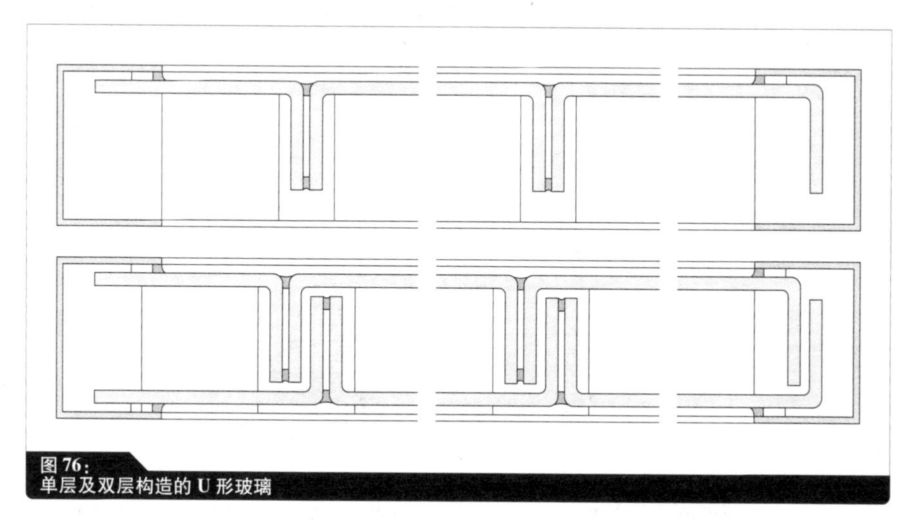

图 76：
单层及双层构造的 U 形玻璃

隔热

只有双层构造才能形成隔热玻璃立面。这种立面是将 U 形玻璃的槽口相对安装的。然后接合处采用定位块及密封材料。对双层构造的立面，应有相应的隔热型框架。〉见图 77

由于 U 形玻璃槽口内的空腔不能像中空玻璃那样做除湿处理，所以必须设计开口让湿空气散出去，以防出现冷凝。U 形玻璃可以是热处理型的，这会改善其结构性能及安全性能。热处理型的 U 形玻璃最长可达约 7m。制造 U 形玻璃时也可辊轧入丝网以防破碎。但这种 U 形玻璃不能做热处理。

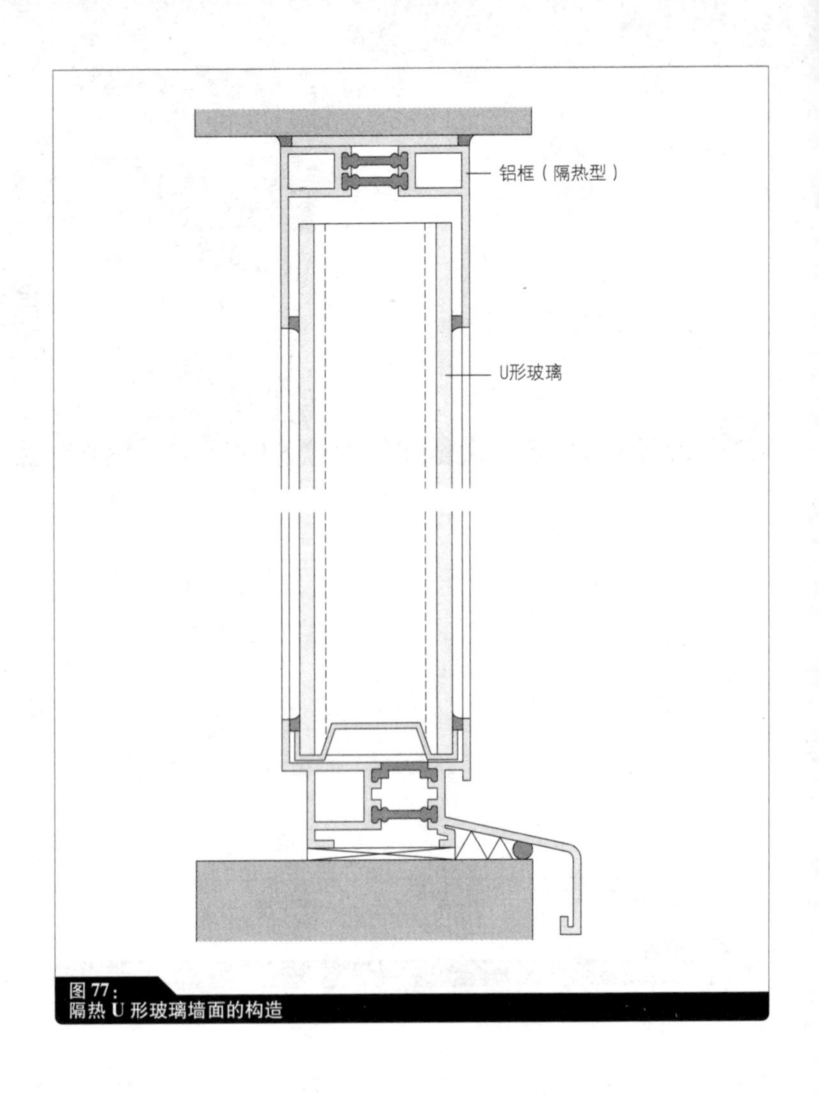


图 77：
隔热 U 形玻璃墙面的构造

表面设计，
镀膜

多种表面的设计都可以用于 U 形玻璃，从光面的到不同图案的。不过由于制作工艺的原因，光面 U 形玻璃的表面不会像浮法玻璃那样平整，这也就降低了它的透明度。〉见图 80 现在，U 形玻璃也可以做成镀膜的，如隔热膜或阳光控制膜。在双层 U 形玻璃面的构造中，有隔热膜的 U 形玻璃应放在内侧，而有阳光控制膜的玻璃用于外侧。还可以将透明隔热材料组合到 U 形玻璃中去。〉见图 78，图 79

图 78：
带有透明隔热层的 U 形玻璃

图 79：
使用 U 形玻璃的立面实例

具体做法是将微孔材料夹进两侧玻璃之间，使之漫射阳光。这种应用主要是为了满足体育馆、工作场所、博物馆、画室之类的建筑对均匀的、无眩光的采光的需要。

P73

**玻璃支承结构**

这种非物质感的结构形式出现在当代，体现出几十年来将玻璃作为建材使用所取得的技术成果。玻璃在建筑中不再局限于扮演表皮的角色，而是开始担当起支承的作用。前面已经介绍了开放型上人采光顶，不过其下部结构一般不是玻璃的。但从 20 世纪 90 年代以来，很多玻璃建筑都使用玻璃作为主要的支承结构材料。这意味着，使玻璃墙面坚固并抵御风荷载的支撑或立柱也是玻璃的。

平板玻璃梁

这种支承结构不是型钢或木制型材，而是纤细的玻璃肋，如夹层安全玻璃等。和立面一样，屋顶的支承构件也可以是玻璃做的。这保证了在有顶庭院或室内可以获得最大程度的通透性及采光。〉见图 81 ~ 图 83

构造形式的日新月异带来全玻璃建筑数目的与日俱增：有玻璃天桥、全玻璃楼梯、实验性的玻璃管支撑体以及拱形与壳形支承系统等。〉见图 84 ~ 图 87

图80：
透明U形玻璃

图81：
框支承玻璃墙面的玻璃支承结构

图82：
点支承玻璃墙面的玻璃支承结构

图83：
格栅式玻璃支撑

承重玻璃管

用硼硅玻璃制成的玻璃管，其截面的结构性能优越，非常适合抵抗高强度的压力。可以用于替代建筑中的混凝土、钢或木的支撑。应特别注意玻璃管端部的钢构件，这部分必须能够将压力均匀传递到玻璃管截面上。荷载通过球性节点传向建筑的相邻部分，球形节点可以防止将剪力和弯矩传到玻璃横截面上。

壳形及拱形支承结构

特别是在主要受压而非受弯的支承结构，如拱、穹顶或壳形支承结构中，才能充分体现玻璃这种抗压强度远高于抗拉强度的性能。

玻璃支承结构的计算

我们所谈论的这些技术进步是建立在以下前提之上的：首先，是将平板玻璃制成钢化、半钢化玻璃或多层构造的夹层安全玻璃的新工艺，这大大提高了新构造的承载力及残余承载力。其次，最新的构造方法的进步得益于在经验和计算基础上得出的对玻璃承载情况的预估。钢之类的柔性材料可以通过塑性变形吸收力；而玻璃不同，应力

图 84：
两个建筑之间的玻璃人行天桥

图 85：
玻璃人行天桥的细部

图 86：
粘结式玻璃楼梯构造

图 87：
粘结式玻璃楼梯的构造细部

集中点会使玻璃突然破碎。某片特定玻璃的荷载极限很难确定，因为这取决于其已受破损（划痕、边缘的小破损）的程度。因此，允许承载力往往会比实际的玻璃荷载极限大为降低。这充分保证了玻璃不会突然破碎，但这也意味着，全玻璃的构造不可能像理论上的那样精巧。

连接件

全玻璃结构的特殊困难之处往往不在于力在单片玻璃内的传递，而在于力在相邻构件间的传递。在这些传力点的范围内，节点形式或为经典的机械式连接（点式或夹式固定），或为粘结节点。粘结连接有将荷载均匀地传给玻璃的优点（因此很适合玻璃的材料特性）。不过，荷载向粘结面的传递会因温湿度、老化等的外部影响减弱。因此，尽管当前的科研正在转向诸如热熔金属箔等强度更高的其他胶粘剂，但在实际中主要还是使用硅胶。

玻璃梁使用的是夹层玻璃，一般由 3 层或更多层的钢化或半钢化玻璃组成。外层的玻璃片提供保护，内层的起实际支承的作用。目前，并无针对此类构造的通行规范，因此不同国家的相关建筑条例有很大的差异。

全玻璃结构的技术极限不仅仅取决于材料强度，还取决于生产的可能性。大多数加工厂可以生产最长 7m、最厚 80mm 的夹层安全玻璃。更长的产品则需要能处理特殊规格产品的高压釜，而一般的公司都没有这种设备。

P77

# 结语

玻璃对建筑师有着巨大魅力，早在布鲁诺·陶特（“天空之家，Haus Des Himmels”，1920）或密斯·凡·德·罗（玻璃摩天楼项目，柏林，1921）等人预见性的作品中就已表明了这点。近90年过去了，这种现代主义的建筑材料仍未失去其先锋地位。

由于气候变化，如今建筑师担负了比以往更大的生态责任。至少随着建筑节能要求的提高，选择有效控制热量及阳光的玻璃种类成了创造热控制玻璃表皮的关键问题。借助于模拟现实软件对建筑影响气候情况的分析，以及从已有建筑中获取的经验，玻璃墙面在节能理念中有了一席之地。近年来，设计领域中也有了新的形式理念。非物质化和通透不再是玻璃建筑中的惟一主导。以彩色玻璃和半透明玻璃形式表现的“物质化”成了流行的设计手法，并有多样化的玻璃产品可供利用。

在未来，我们极有可能看到可调式构造层、中空玻璃及自洁表面之类的新技术愈发重要。今天显而易见的是，玻璃的技术发展方兴未艾。

在这种建材的应用和设计领域里我们一路不断发现新的风景，未来，它的潜力仍会留给我们一片新奇有趣的天地。

P78

# 附录

P78

## 标准、导则与规范

**建材玻璃**

| | |
|---|---|
| EN 572 | 建材玻璃—普通钠钙硅酸盐玻璃 |
| EN 1051 | 建材玻璃—玻璃砖及玻璃面砖 |
| EN 1096 | 建材玻璃—镀膜玻璃 |
| EN 1279，1~6 部分 | 建材玻璃—中空玻璃组件 |
| EN 1863 | 建材玻璃—半钢化钠钙硅酸盐玻璃 |
| EN 12150 | 建材玻璃—钢化钠钙硅酸盐玻璃 |
| EN 14179 | 建材玻璃—均质钢化玻璃 |
| EN ISO 12543 | 建材玻璃—夹层玻璃及夹层安全玻璃 |
| ASTM C1048 - 04 | 半钢化平板玻璃标准 - HS 类、FT 类镀膜及非镀膜玻璃 |
| ASTM C1172 - 03 | 建筑夹层平板玻璃标准 |
| ASTM C1376 - 03 | 高温及真空镀膜平板玻璃标准 |
| ASTM C1464 - 06 | 曲面玻璃标准 |

**隔热**

| | |
|---|---|
| EN 673 | 建材玻璃—传热限定（*U* 值） |

**遮阳**

| | |
|---|---|
| EN 410 | 建材玻璃—玻璃光学及采光性能限定 |
| ISO 9050：2003 | 建材玻璃—对玻璃的透光性、阳光直接透射、总太阳能透过量、紫外线透射及相关参数的限定 |

**安全**

| | |
|---|---|
| EN 356 | 建材玻璃—防暴玻璃—防人为袭击的测试及等级 |
| EN 1063 | 建材玻璃—防暴玻璃—防子弹射击的测试及等级 |

**耐火**

| | |
|---|---|
| EN 357 | 建材玻璃—透明或半透明玻璃制品制成的防火玻璃组件—耐火等级 |
| EN 13501 - 1 | 建筑产品及组件的耐火等级—第一部分：依据火灾实验数据的等级划分 |

| 稳定性、耐久性 | |
|---|---|
| EN 13022 | 建材玻璃—结构密封胶粘结玻璃 |
| EN 13474 | 建材玻璃—面板玻璃设计 |
| ASTM E2358 - 04 | 非临时性玻璃栏杆、栏板及防护装置的玻璃性能标准 |
| ANSI Z97. 1 - 2004 | 美国国家标准—建筑安全玻璃材料—安全性能标准及测试方法 |
| EOTA | 系统的欧洲技术认定导则（SSGS） |

P80

## 参考书目

Achilles, Andreas. "Coloured Glass: Manufacture, Processing, Planning." *Detail: Review of Architecture and Construction Detail* 2 (2007): 184–87.

Button, David, and Brian Pye, eds. *Glass in Building*. Oxford 1993.

Compagno, Andrea. *Intelligente Glasfassaden/Intelligent Glass Façades: Material Anwendung Gestaltung/Material, Practice, Design*. 5th ed. Basel: Birkhäuser Verlag, 2002.

Krippner, Roland, and Florian Musso. *Basics Facade Apertures*. Basel: Birkhäuser, 2008.

Kruft, Hanno-Walter. *A History of Architectural Theory: From Vitruvius to the Present*. London: Zwemmer; New York: Princeton Architectural Press, 1994.

Rice, Peter, and Hugh Dutton. *Structural Glass*, London; New York: E & FN Spon, 1995.

Schittich, Christian, ed. *Building Skins*. Munich: Edition Detail; Basel: Birkhäuser, 2006.

Sobek, Werner. "Glass Structures." *The Structural Engineer* 83/7 (April 2005).

Staib, Gerald, Dieter Balkow, Matthias Schuler, and Werner Sobek. *Glass Construction Manual*. Munich: Edition Detail, 2006.

Weller, Bernhard, and Thomas Schadow. "Structural Use of Glass." *Detail: Review of Architecture and Construction* 2 (2007): 188–90.

Wurm, Jan. *Glass Structures: Design and Construction of Self-Supporting Skins*. Basel: Birkhäuser Verlag, 2007.

P81 **图片版权**

Figs. 2, 4, 8, 17, 18, 19, 20, 21, 22, 29, 31: photographer: Martin Lutz (Akademie der Bildenden Künste), Stuttgart; copyright: Andreas Achilles, Jürgen Braun, Peter Seger, Thomas Stark, Tina Volz, Stuttgart

Fig. 3, figs. 24, 25 (Cineplexx Salzburg); figs. 26, 27, 78; fig. 79 (IHK Würzburg); fig. 80 (Uni-Klinik Hamburg): Glasfabrik Lamberts GmbH & Co. KG, Wunsiedel

Fig. 23: Saint-Gobain Glass Deutschland GmbH, Aachen

Fig. 28 (Landeszentralbank Meiningen): Schott AG, Mainz

Fig. 30 (Technologie und Innovationszentrum Grieskirchen); fig. 66 (Douglasgebäude Linz); fig. 82 (Palais Coburg, Vienna): Eckelt Glas GmbH, Steyr

Fig. 36: Schüco International KG, Bielefeld

Figs. 37, 38, 40, 41, 42: Institut für internationale Architektur-Dokumentation GmbH & Co. KG, Redaktion Detail, Munich

Fig. 44 (Kunsthaus Bregenz): Hélène Binet, London

Fig. 59: Siegfried Irion, Stuttgart

Fig. 83 (Technische Universität Dresden): Institut für Baukonstruktion, Technische Universität Dresden, Dresden

Figs. 84, 85 (Glass bridge, Schwäbisch Hall): Glas Trösch Beratungs-GmbH, Ulm-Donautal

Fig. 86 (Glasstec Düsseldorf): René Tillmann, Düsseldorf

Fig. 87: Andreas Fuchs, Universität Stuttgart IBK Forschung und Entwicklung, Stuttgart

Figs. 1, 5, 6, 7, 9, 10, 11, 12, 13, 14, 15, 16, 32, 33, 34, 35; fig. 39 (LBBW high-rise, Stuttgart); fig. 43 (Kronen Carré, Stuttgart); figs. 45, 46, 47; fig. 48 (office building on Königstraße Stuttgart); figs. 49, 50, 51, 52, 53, 54, 55, 56, 57, 58, 60, 61, 62, 63, 64, 65, 67, 68, 69, 70, 71, 72, 73, 74, 75, 76, 77; fig. 81 (Kunstgalerie Stuttgart): Andreas Achilles, Diane Navratil, Stuttgart

P81 **作者简介**

安德烈斯·艾奇里斯（Andreas Achilles），工程师，曾任德国斯图加特大学建筑工程与艺术学院（Baukonstruktion und Entwerfen）讲师，现为自由执业建筑师及作家，工作地点为斯图加特。

黛安·娜维拉蒂尔（Diane Navratil），工程师（建筑与城市规划专业），供职于德国卡尔斯鲁厄城市规划部门。